97-REM-5

ASSESSING BIOAVAILABILITY OF METALS IN BIOSOLIDS-TREATED SOILS

by

A.C. Chang, Ph.D., P.E.
D.E. Crowley, Ph.D.
A.L. Page, Ph.D.

Department of Environmental Sciences
University of California, Riverside

2003

Water Environment Research Foundation

The Water Environment Research Foundation, a not-for-profit organization, funds and manages water quality research for its subscribers through a diverse public-private partnership between municipal utilities, corporations, academia, industry, and the federal government. WERF subscribers include municipal and regional water and wastewater utilities, industrial corporations, environmental engineering firms, and others that share a commitment to cost-effective water quality solutions. WERF is dedicated to advancing science and technology addressing water quality issues as they impact water resources, the atmosphere, the lands, and quality of life.

For more information, contact:
Water Environment Research Foundation
635 Slaters Lane, Suite 300
Alexandria, VA 22314-1177
Tel: (703) 684-2470
Fax: (703) 299-0742
www.werf.org
werf@werf.org

This report was co-published by the following organization. For nonsubscriber sales information, contact:

IWA Publishing
Alliance House, 12 Caxton Street
London SW1H 0QS, United Kingdom
Tel: +44 (0) 20 7654 5500
Fax: +44 (0) 20 7654 5555
www.iwapublishing.com
publications@iwap.co.uk

Library of Congress Catalog Card Number: 2003109624

IWAP ISBN: 1-84339-679-3

This report was prepared by the organization(s) named below as an account of work sponsored by the Water Environment Research Foundation (WERF). Neither WERF, members of WERF, the organization(s) named below, nor any person acting on their behalf: (a) makes any warranty, express or implied, with respect to the use of any information, apparatus, method, or process disclosed in this report or that such use may not infringe on privately owned rights; or (b) assumes any liabilities with respect to the use of, or for damages resulting from the use of, any information, apparatus, method, or process disclosed in this report.

University of California, Riverside

This document was reviewed by a panel of independent experts selected by WERF. Mention of trade names or commercial products does not constitute WERF nor EPA endorsement or recommendations for use. Similarly, omission of products or trade names indicates nothing concerning WERF's nor EPA's positions regarding product effectiveness or applicability.

The research on which this report is based was funded, in part, by the United States Environmental Protection Agency through Cooperative Agreement No. CR-825237-01 with the Water Environment Research Foundation (WERF). Unless an EPA logo appears on the cover, this report is a publication of WERF, not EPA. Funds awarded under the Cooperative Agreement cited above were not used for editorial services, reproduction, printing, or distribution.

ACKNOWLEDGMENTS

Investigators of the study gratefully acknowledge Metropolitan Water Reclamation District of Greater Chicago, City of Riverside Regional Water Pollution Control Plant, and City of Los Angeles Hyperion Wastewater Treatment Plant for providing biosolids that were included in the experiments. Dr. Robert H. Dowdy, University of Minnesota, St. Paul, MN and Dr. Thomas C. Granato, Metropolitan Water Reclamation District of Greater Chicago provided data and soil and biosolids samples collected from long-term biosolids land applications sites for the field validation of the study.

Report Preparation

Principal Investigators:
Andrew C. Chang, Ph.D., P.E.
David E. Crowley, Ph.D.
Albert L. Page, Ph.D.
Department of Environmental Sciences, University of California, Riverside

Project Team:
Bon-Jun Koo, Ph.D.
Carl Nelson, M.S.
David Thomason, A.A.
Natalie Krage, M.S.
Department of Environmental Sciences, University of California, Riverside

Project Subcommittee

Patrick L. Brezonik, Ph.D.
Department of Civil Engineering
University of Minnesota

Jane B. Forste
Jane Forste Assocoiates, LLC

Thomas Granato, Ph.D.
Metropolitan Water Reclamation District of Greater Chicago

Robert E. Pepperman, Ph.D.
United Water Service

Water Environment Research Foundation Staff

Director of Research: Daniel M. Woltering, Ph.D.
Project Manager: Lola Olabode, M.P.H.

ABSTRACT AND BENEFITS

Abstract:

Plant-availability of metals in biosolids-treated soils may be mathematically described by $M_t = C\,(1 - e^{-kt})$ where M_t (mg kg^{-1}) is the cumulative metal removal from the biosolids-treated soils by growing and harvesting plants for t years, C is the total phytoavailable metal pool of the soil (mg kg^{-1}) at $t = 0$, and k is the metal absorption rate coefficient (yr^{-1}). The total available metal pool (C) is defined as metals extractable by organic acids in the rhizosphere of growing plants and k is related to the kinetics of metal release by organic acids. Half-life of the available metals in biosolids-amended soils may be derived from k. Experiments were conducted to characterize the concentration and composition of the organic acids. A successive extraction method was used to extract metals from biosolids-treated soils for determining C and k. In this manner, the plant available metals of the biosolids-treated soils are defined by the total available metals, half-life, and duration of plant growth. Examples showed that the total available metals of the biosolids-amended were 18-39, 0.12-2.0, 7.4-13, 24-46, 1.2-2.1, and 43-53% of the total Cd, Cr, Cu, Ni. Pb, and Zn in the soils, respectively.

Benefits:

- Provide biosolids land application planners and practitioners a clear conceptual understanding of how metals in the biosolids-amended soils may be dissolved and absorbed by plants.
- Develop a mathematical expression to describe the plant-availability of metals in biosolids-amended soils.
- Develop a method for evaluating the plant-availability of metals in biosolids-amended soils using laboratory-based soil extractions with organic acids.
- Demonstrate the usefulness of the method through two examples where biosolids were applied for a long period of time.

Keywords: Land application, trace elements, heavy metals, root exudates, rhizosphere, organic acids, plant uptake

TABLE OF CONTENTS

LIST OF TABLES

LIST OF FIGURES

LIST OF ACRONYMS

AAS	Atomic absorption spectrophotometry
ANOVA	Analysis of variance
DTPA	Diethylene triamine penta acetic acid
EDTA	Ethylene diamine tetra acetic acid
GC-MS	Gas chromatography-mass spectroscopy
HPLC	High performance liquid chromatography
HyA-Mt	hydroxyaluminum-montmorillonite
ICP-OES	Inductively coupled plasma-optical emission spectroscopy
MWRDGC	Metropolitan Water Reclamation District of Greater Chicago
NMR	Nuclear magnetic resonance
POTW	Publicly owned treatment works
U.S. EPA	U.S. Environmental Protection Agency

EXECUTIVE SUMMARY

The phytoavailability of metals in biosolids-treated soils may be described by the following model:

$$M = C\,(1 - e^{-k\,t})$$

where M (mg kg^{-1}) is the cumulative metal removal from the biosolids-treated soils by growing and harvesting plants for t years, C is the total phytoavailable metal pool (mg kg^{-1}), and k is the metal release coefficient (yr^{-1}) of the soils. If the C and k of a soil are defined, the phytoavailability of the metals in the biosolids receiving soils may be evaluated.

Organic acids present in the rhizosphere of growing plants are widely recognized to be responsible for dissolving the solid phase metals in the soil and making them available for plant absorption. Studies were initiated to investigate the concentration and composition of organic acids in the biosolids-treated plant growing media and the role of rhizosphere organic acids on the solubilization and phytoavailability of biosolids-borne metals. Based on the outcomes of the investigations, synthetic organic acid mixtures were formulated to assess the dissolution of metals from various biosolids sources in the biosolids-treated soils. Finally, soil and plant tissue samples were obtained from two long term biosolids land application sites to validate the metal phytoavailability model for the biosolids-treated soils with field data.

The findings are summarized as follows:

1. **Literature Review** – The nature of root exudates and their role in plant nutrition and soil chemistry has been studied for over two decades. Much of the root exudate literature consisted of results from growing plant seedlings for up to 14 days under aseptic conditions. Studies have demonstrated that the primary component of the root exudates are low molecular weight organic acids which might enhance the availability of sparingly soluble soil Fe, P, and Zn to growing plants. Microbial metabolism may play an important role in modifying the composition of organic substances found in the rhizosphere. The composition of the organic acids in the root exudates varied considerably between the plant species studied. Even for the same plant species, investigators reported the presence of different chemical components.

   Plant Growth - The biomass harvests of plants that were cultured with the biosolids-treated media were equal or better than those cultured with the standard plant nutrient solution media. The growth of plants was not adversely affected and exhibited no visual symptom of injuries even when the metal concentrations in plant tissue were significantly higher than normally observed. The increases of the biomass production for plants that were associated with biosolids treatment were consistent over the entire experimental period of 16 weeks, plant species tested (e.g., corn, wheat, Sudan grass, Swiss chard, canola, and chickpea), and biosolids sources (e.g., MWRDGC 1975, Milorganite, MWRDGC 1999, Los Angeles, and Riverside). The increase of biomass yields for plants grown on biosolids-treated media over the standard nutrient solution was attributed to the presence of biosolids in the media, which increased the intensity of availability of nutrients in the rhizosphere.

3. **Root Exudates in Biosolids-treated Media** – Organic acids were the primary substances found in root exudates that were recovered from all of the plant-growth media. Even the without-plant and the without-biosolids blanks produced organic acids. The biosolids treatment and the presence of plants enhanced the volume of organic acids produced. When plants were present in biosolids-treated media, the production of organic acids was further enhanced. Acetic, lactic, butyric, and oxalic acids were the primary components of organic acids found in the plant-growth media. Collectively, they accounted for 75-90% of the total organic acids recovered. Other organic acids detected were propionic, pyruvic, succinic, tartaric, maleic, and glutaric acids. The organic acid composition of the plant growing media did not change significantly with plant species, varieties of the same species, duration of plant growth, and sources of biosolids. The composition of recovered root exudates was independently confirmed by using GC-MS and the NMR methods. The concentrations of organic acids in the plant growing media were estimated to be 0.001 to 0.1 M of COO^-, depending on the water content of the media. Based on the compositional analysis, synthetic organic acid mixtures comparable to the composition of root exudates recovered from the rhizosphere of biosolids-treated media were formulated to evaluate the dissolution of metals from biosolids-treated soils.

4. **Metal Uptake by Plants** – The concentrations of metals in the plants grown on biosolids-treated media were significantly higher than those of the same plants grown on the standard plant nutrient media. The amounts of metal transferred from the biosolids-treated growing media to the plants varied with the metal elements. Following in a descending order, the amounts transferred were: Cd > Ni = Zn > Cu = Pb > Cr. For corn grown on biosolids-treated media from seedlings to maturity, the amounts of metals absorbed by the plants were 14, 0.01, 1.6, 9.8, 1.4, and 8.6% of the biosolids-borne Cd, Cr, Cu, Ni, Pb, and Zn in the rhizosphere, respectively. For Cd, Cu, Ni, and Zn, the differences in metal uptake between plant species and varieties were relatively minor in comparison to the differences of metal uptake between plants grown on the standard plant nutrient solution- and the biosolids-treated media. The rate of metal uptake during the course of plant development may be calculated as the amounts of metal accumulated in plant tissue per unit biomass per unit growing period. The rate of Cd and Zn uptake for corn shoots remained rather constant for the first 8-10 weeks, metabolically the most active period of the growing season. Toward the end of the growing season, the uptake rates tended to rise as the biomass stopped accumulating while the uptake of water and metals continued. For Pb, the uptake rate by corn shoots grown on biosolids-treated media decreased with the increasing length of growing

period. This indicated that the solution concentration of Pb, and thus its availability to the plant, is controlled by the rate of dissolution of biosolids-borne Pb.

5. **Metal Solubility in Organic Acids** – The solubility of metals in a biosolids-treated soil with 0.01 and 0.1 M synthetic organic acid mixtures was significantly higher than that at 0.001 M. The slight variations in the organic acid compositions due to biosolids treatment, length of growing periods, and plant species did not affect the solubility of the metals. The conclusion was reached by first using the Student-Newman-Keuls test to compare amounts of metals extracted by standard media vs. biosolids-treated media. The differences were not significantly different at $p < 0.05$. The data were then pooled to compare the difference between plant growing periods using ANOVA. Again, the differences were not significant at $p < 0.05$. Results of this experiment indicated that the concentrations of synthetic organic acid mixtures significantly affected the amounts of metals extracted from both the control and the biosolids-treated soils.

6. **Metal Dissolution Kinetics**. Results of the experiments indicated that a rapid dissolution of metals occurred in the first 15 minutes of the equilibration. For Cd, Cu, Ni, and Zn, approximately 60-70% of the extractable metals were released in the first 15 minutes. For Cr and Pb, the initial releases were approximately 30% of the total. Following the initial release, a slow but steady release of metals occurred over the next 48 hours. Based on results of the metal dissolution studies, the synthetic organic acid mixture of 0.01 M COO^-, 13.5 mM $Ca(NO_3)_2$ background matrix, and pH 4.8 was selected as the metal extractant for the field validation.

7. **Field validation**. Three biosolids application fields at the Fulton County Biosolids Land Reclamation Site of the Metropolitan Water Reclamation District of Greater Chicago and 16 experimental plots of long-term biosolids land application at the University of Minnesota were employed in field validation.

   The biosolids-treated soils from the testing plots were successively extracted with the synthetic organic acids for 10 consecutive times. The data showed that the cumulative metal released by the organic acids in the synthetic root exudates with respect to the volume of organic acids used followed the first order reaction model that:

$$M_q = a\,(1 - e^{-bq})$$

   where M_q is the cumulative metal removal from biosolids-treated soils (mg kg^{-1}) by q volume of organic acids (mole COO^- kg^{-1}), a is the total synthetic root exudates extractable metal (mg kg^{-1}), and b is the synthetic root exudates metal release rate constant. The parameter a in this equation corresponds to the total phytoavailable metal pool (C) in the metal phytoavailability model.

The field sampling-derived plant uptake data, along with the estimates of the total phytoavailable metals pool of the soils, then were used to fit the metal phytoavailability model to obtain the field-based metal plant uptake constant, *k*. Let A equal the effective organic acids production factor of the biosolids-treated soils. The *k* of a biosolids-treated soil may be mathematically related to the synthetic root exudates metal release constant of the corresponding soils as $k = b/A$, where A followed a normal distribution with a mean of 3.8 and standard deviation of 1.2.

CHAPTER 1.0

INTRODUCTION

In cropland application, biosolids are potentially harmful because they contain disease-causing organisms and/or chemical pollutants. In the soil, pathogens rapidly die off and do not present long-lasting detrimental environmental effects. Chemical pollutants such as metals and metalloids, however, may persist in receiving soils and be absorbed in sufficient quantities by growing plants to adversely affect the health of consumers and/or plants. In 1972, Congress directed the U.S. Environmental Protection Agency (U.S. EPA), through Section 405(a) of the Federal Water Pollution Control Act, to develop guidelines for the use and disposal of biosolids, including cropland application. Regulations for the use or disposal of sewage sludge (at CFR 40, Parts 257, 403, and 503) were finalized on February 19, 1993 (U.S. EPA, 1993).

The goal of the regulations is to encourage beneficial use of biosolids and to protect public health and the environment from any "reasonably anticipated adverse effect" of pollutants found in biosolids. These regulations define the domain within which biosolids may be safely disposed or beneficially used. They have far-reaching implications on how biosolids generated by more than15,000 publicly owned treatment works (POTW) nationwide are managed, and provide the POTWs with guidelines to develop long-range, cost-effective biosolids management plans. With the regulations in place, cropland application has become a viable option for the management of biosolids. It has been estimated that 55% of the nation's annual biosolids production (6.85×10^6 tons yr^{-1}) is now land applied (Bastian, 1997). Over 60% of the POTWs in California, including those serving the San Francisco, Los Angeles, and San Diego metropolitan areas, employed land application in their biosolids management plans (California Association of Sanitation Agencies, 1999).

The numerical limits in current regulations are based on scientific data accumulated in the past two decades. Are U.S. EPA regulations protective over the long-term with respect to accumulation of potentially toxic metals and metalloids from cropland application of biosolids? To address this question, the phytoavailability of metals in biosolids-treated soils must be objectively evaluated.

1.1 Rationale for Root Exudate-based Metal Phytoavailability

Plants absorb potentially toxic metals that are dissolved in soil solution. Because potentially toxic metals exist in the natural environment primarily in sparingly soluble solid phases, the reactivity and phytoavailability of metals in terrestrial and aquatic environments are dependent on the chemical equilibria between those in solution and those associated with solid phases as well as by the kinetics of these reactions (Sposito, 1981; Gerritse et al., 1983; Corey et al., 1987a). When conditions for plant growth are otherwise the same, activities of the free metal or metalloid ions in the soil solution determine their availability to plants (Sparks, 1984; Parker et al., 1995).

In land application of biosolids, metals are added to the soil mainly in solid phases. In this case, plant absorption of biosolids-borne metals may take place through root extraction from those present in bulk soil solution via solid-solution phase equilibration or in the rhizosphere where root exudates are presumed to mobilize the solid phase metals. Considering that metal concentrations in soil solution are always low and dissolution kinetics for metals in soils are also commonly low, plant absorption from the bulk soil solution may not be the most important process by which metals are absorbed by plants grown in biosolids-treated soils. Therefore, the root absorption of biosolids-borne metals in the soils is most likely dictated by short-range mass transfer process. Candalaria and Chang (1997) demonstrated that, in soils, the biosolids-borne metals remained in their discrete solid phases. Eaton (1936) reported that plants grown in sand culture obtained Fe and P "from water-insoluble minerals by absorption across particle-root interfaces rather than by absorption from culture or soil solution." Jenny and Overstreet (1938) postulated that plants might take up mineral elements through contact and ion interchange.

Root exudates have been found to increase availability of nutrients (Romheld and Marschner, 1986; Fan et al., 1997) and enhance the trace element uptake by plants (Welch, 1995). Mench and Martin (1991) found that soil Cd, Cu, Fe, Mn, Ni, and Zn were solubilized by root exudates of *Nicotiana tabacum* L., *N. rustica* L., and *Zea mays* L. The chemical compositions of root exudates recovered from these three plant species differed. They also reported that Cd concentrations of these three plant species were proportional to amounts of Cd extracted from the soil in which they were grown. Krishnamurti et al. (1997) demonstrated that soil Cd might be mobilized by low-molecular-weight organic acids such as acetic, citric, oxalic, fumaric, and succinic acids, which were often found in root exudates. At the same Cd concentration, the release of Cd from the soil in the presence of organic acids decreased in the order: fumaric acid > citric acid > oxalic acid > acetic acid > succinic acid.

The phytoavailability of a metal in soil customarily is defined by the amount of metal absorbed by growing plants or by concentration in the harvested plant tissue. Plants grown on biosolids-treated soils typically absorb << 1% of the metals present in the soil. It would take a long time to deplete the biosolids-borne metals that are available for plant absorption in biosolids-treated soils. Therefore, phytoavailability of biosolids-borne metals must account for not only the plant uptake in one growing season, but also the total amounts available over time. Neither the plant uptake nor the concentration of metals clearly reflects the labile pool of metal in soil. To define the phytoavailability of metals in the biosolids-receiving soils, the intensity factor (rate at which the biosolids-borne metals are absorbed by plants) and capacity factor (total amounts of phytoavailable metals in biosolids-receiving soils) must be addressed.

We hypothesize that the total amount of phytoavailable metals in biosolids-treated soils may be determined as the total amount of metals soluble in root exudates. The rate of uptake by plants, which determines the plant tissue concentration and phytotoxicity of a metal, is determined by the kinetics of metal mobilized by rhizosphere soil solution (i.e., root exudates plus background matrix of soil solution). In this manner, the phytoavailability of biosolids-borne metals may be defined in terms of a capacity factor (metals solubilized by root exudates), which describes "how much" of the metals is available, and an intensity factor (the rate at which metals may be solubilized), which relates the amounts of metal to be absorbed by plants.

Based on this approach, the availability of biosolids-borne metals in the soil may be expressed in the following manner:

$$M = C\,(e^{-kt}) \qquad \text{[Eq. 1.1-1]}$$

where M is the amounts of metal removed by crops at year t, C is the capacity factor representing the labile metal pool of the biosolids-treated soils, and k is the intensity factor representing the rate metals are released in the rhizosphere and, therefore, become available for plant absorption. Equation 1.1-1 may be written in the following form:

$$M_t = C\,(1-e^{-kt}) \qquad \text{[Eq. 1.1-2]}$$

where M_t (mg kg $^{-1}$) is the cumulative metal removal from the soils that results from growing and harvesting plants for t years (yr), C is the total phytoavailable metal pool (mg kg $^{-1}$), k is the metal release coefficient (yr $^{-1}$) of the soils. If the C and k of a soil are defined, the phytoavailability of the metals in the biosolids amended soils may be evaluated in terms of total plant available metals, amounts released each growing season, and the half-life of metal uptake.

Field data may be used to establish C and k for soils. The dissolution behavior of biosolids-borne metals in the root exudates will define the relationship between C and the total root exudates-extractable metals and between k and the metal dissolution rate constants.

1.2 Study Objectives

The purpose of this investigation is to define the role of root exudates in plant uptake of metals from biosolids-treated soils, and to evaluate the long-term phytoavailability of biosolids-borne metals accumulated in soils.

Specific objectives are to:

- Develop an experimental procedure to collect root exudates of plants grown with and without biosolids using a non-sterile hydroponics sand culture system.
- Characterize the chemical compositions of root exudates of selected plant species.
- Evaluate the effects of plant species, soil and biosolids properties, and time on the chemical composition of root exudates.
- Formulate synthetic root exudates that can be used as surrogates for actual root exudates to study the chemistry of metals in biosolids-treated soils.
- Evaluate the role of root exudates in solubilizing metals in biosolids-treated soils by determining root exudate-specific metal solubility and dissolution rate constants of biosolids-treated soils.
- Correlate metal concentrations of plants grown on biosolids-treated soils with root exudate-specific metal dissolution rate coefficients for these soils.

CHAPTER 2.0

METHODS AND PROCEDURES

The study included a comprehensive review of the root exudates related literature. Based on the findings of the literature review, the experimental procedures and methods for recovering and analysis of root exudates were determined. The results of the literature review are summarized in Chapter 3.0.

Experiments were set up in greenhouses where plants were grown in nutrient solution irrigated sand culture media and biosolids-treated sand culture media. For the biosolids treatments, the Zn, Cu, Ni, and Mn supplements of the nutrient solution were withheld and the plants absorbed the trace elements from the added biosolids. Plants were cultivated under septic conditions and no attempt was made to control the microbial population in the growing media. At the end of the growing periods, the biomass was harvested and organic acids in the root zone were recovered for chemical assays. In addition, organic acids were collected from the sand media that received the same nutrient solution and biosolids treatments but were not cultivated. These are designated hereafter as the nutrient solution and biosolids blank treatments that were included with each set of experiments.

Experiments conducted are summarized as follows:

- Comparing growth, metal uptake, and root exudate composition of corn (*Zea mays* L.) grown on biosolids-treated and standard plant nutrient solution media for 2-16 weeks.
- Comparing growth, metal uptake and root exudate composition of six species of plants [corn (*Zea mays* L.), Durum wheat (*Triticum turgidum* var. durum L.), canola (*Brassica napas* L.), Sudan grass (*Sorghum sudanense*), chickpea (*Cicer arietinum* L.), and Swiss chard (*Beta vulgaris* L.)] grown on biosolids-treated sand and standard plant nutrient solution media.
- Comparing growth, metal uptake and root exudate composition of three varieties of wheat (*Triticum aesticum* L. cv. Durum, *Triticum turgidum* L. cv. Yecora Rojo, and *Triticum turgidum* L. Chinese spring) grown on biosolids-treated and standard plant nutrient solution media.
- Comparing growth, metal uptake and root exudate composition of corn grown on culture media treated with biosolids from five different sources (Milorganite; biosolids produced by MWRDGC in 1975; and dewatered biosolids produced by MWRDGC in 1999, City of Los Angeles, California, and City of Riverside, California). All biosolids were collected in 1999, except for the biosolids of MWRDGC in 1975. They are referred hereafter as

Milorganite, MWRDGC 1975, MWRDGC 1999, Los Angeles, and Riverside, respectively.

- ♦ Solubility and dissolution kinetics of metals of biosolids-treated soils in synthetic root exudates. Procedures for this part of the study are described in Section 2.2.

The data generated from the experiments then were coupled with data collected from field validation experiments, to define parameters in the following metal availability model for biosolids treated soils such that:

$$M_t = C\,(1 - e^{-kt}) \qquad \text{[Eq. 1.1-2]}$$

where M_t is the metal removed by crops in t years (i.e., growing seasons), C is the capacity factor that defines the labile metal pool of the biosolids-treated soils, and k is the intensity factor that defines the rate metal in the rhizosphere of the biosolids-treated soils is available to plants.

The general procedures used for these experiments are summarized as follows.

2.1 Greenhouse Experiments

Each plant-culturing unit consisted of a reservoir for nutrient solution and four 10-liter plastic pots situated immediately above the nutrient reservoir. The plastic pots had a standpipe to control the water level during irrigation and perforations at the bottom for drainage. Each pot contained approximately 11.5 kg of silica sand to support plant roots. For the biosolids treatment, an archived biosolids compost produced by the Metropolitan Water Reclamation District of Greater Chicago in 1975 (MWRDGC 1975, hereafter) was used at a rate equivalent to 20 Mg ha^{-1} in all experiments. Concentrations of some metals in the compost exceeded the maximum contaminant level (MCL) for biosolids of exceptional quality, as specified in 40 CFR 503 (Table 2-1). Compared to present day biosolids, the metal concentrations of this material represented an extreme upper boundary for municipal biosolids. Experiments also were conducted to compare responses of contemporary biosolids against MWRDGC 1975.

Table 2-1. Concentration of Selected Trace Metal Elements in MWRDGC 1975 Biosolids.

Element	Concentration (mg kg^{-1})		
	Range	Mean	Standard Deviation
As	20 – 25	23	1
Cd	204 – 250	229	12
Cr	3,100 – 3,800	3,480	202
Cu	1,030 – 1,460	1,330	92
Ni	656 – 771	721	31
Se	2.0 – 3.8	2.8	0.3
Zn	4,260 – 5,090	4,800	274

Data derived from Sommers et al., 1991.

Recently produced biosolids from four other sources, namely Milwaukee, Wisconsin (Milorganite, hereafter), Metropolitan Water Reclamation District of Greater Chicago (MWRDGC 1999, hereafter), City of Los Angeles (Los Angeles, hereafter), and City of Riverside, California (Riverside, hereafter), were also used in experiments. The concentrations of Cd, Cr, Cu, Ni, Pb, and Zn in these biosolids were considerably lower than those in the MWRDGC 1975 (Table 2-2).

Table 2-2. Nitrogen Content and Concentrations of Selected Metals in Biosolids from Five Sources.

Source of Biosolids[a]	Total N (%)	Metal Element (mg kg^{-1})					
		Cd	Cr	Cu	Pb	Ni	Zn
MWRDGC 1975	0.8	119	1556	767	1344	482	4390
MWRDGC 1999	1.6	7	112	390	248	56	1709
Milorganite	5.8	5	210	196	161	57	1476
Los Angeles	4.4	21	94	789	116	102	1810
Riverside	3.4	3	56	630	131	41	1728

[a]MWRDGC 1975: air-dried Imhoff tank sludge, MWRDGC 1999: air-dried dewatered anaerobically digested biosolids, Milorganite: packaged product (derived from heat-dried biosolids of Milwaukee, WI) purchased locally, Los Angeles: dewatered anaerobically digested biosolids, and Riverside: air-dried dewatered anaerobically digested biosolids.

Seeds were germinated in a growth chamber and seedlings were transferred to pots 3-5 days after the onset of germination. Plants were irrigated 4-5 times daily by nutrient solution in the reservoir through an automated pumping system. The standard nutrient solution contained 2 M $Ca(NO_3)_2$, 1 M KCl, 0.5 M $MgSO_4$, 0.1 M NaCl, 1 M $C_6H_{13}NSO_4$ buffer, 20 x 10^{-3} M NaH_2PO_4, 12 x 10^{-3} M Zn-EDTA, 150 x 10^{-3} M Fe-EDTA, 0.6 x 10^{-3} M Mn-EDTA, 2 x 10^{-3} M Cu-EDTA, 0.1 x 10^{-3} M Ni-EDTA, 10 x 10^{-3} M H_3BO_3, and 0.1 x 10^{-3} M Na_2MoO_4. The EDTA-chelated trace elements in the nutrient solution were produced by combining chloride or nitrate salts of the elements with corresponding amounts of H_2-EDTA. For the biosolids treatments, Zn, Mn, Cu, Ni, and H_2-EDTA were withheld from the nutrient solution. In this manner, metals absorbed by plants receiving the biosolids treatment came exclusively from the biosolids amendments. All of the pots supported by an individual reservoir received the same treatments (i.e., plant species, plant growing period, and biosolids treatment). After irrigation, excess nutrient solution drained into the reservoir by gravity. One hundred liters of nutrient solution were used for each reservoir

2.1.1 Root Exudate Extraction

For root exudate collection, plants were grown under septic conditions because plant roots naturally have a high microbial density associated with the rhizosphere. Root exudates accumulating under a septic environment depicted the plant growing conditions in the field more realistically. To collect root exudates, pots were removed from the plant-growing unit, the sand media were flushed three times with deionized water, and the plants were allowed to photosynthesize for five hours. Afterwards, the root exudates were eluted with two liters of deionized water, collected in plastic bottles and stored in a freezer at –80°C until the time for concentrating.

2.1.2 Biomass and Plant Tissue Samples

Following the collection of root exudates, plants were removed from the pots and separated into shoot and root portions, cleaned with deionized water, dried at 65°C for five days, and weighed for biomass. Aliquots of plant tissues were digested in Teflon Parr bombs by a

HNO_3 microwave digestion procedure for metal determination using ICP-OES spectroscopy (Milward and Klukner, 1989).

2.1.3 Concentration and Analysis of Root Exudates

The amounts of root exudates collected from the plant growing media were small relative to the volume of water used for flushing. For chemical analysis, 250 mL of the stored root exudate solution was freeze-dried. Once the root exudates were lyophilized, they were stored in a –20°C freezer until they were ready for chemical analysis. Organic acids in the concentrated root exudate aliquots were analyzed. Results of past investigations indicated that organic acids are the primary constituents of root exudates and also are chemically the most active component of root exudates for reacting with metals. We employed a Dionex Anion Chromatographic Instrument with an IonPac ICE-AS11 column to determine the organic acid composition of the root exudates. For selected samples, organic acids identified by the anion chromatography were confirmed by NMR and by GC-MS methods.

2.2 Laboratory Experiments

We set up laboratory experiments to test whether the differences in organic acid compositions and concentrations affected the solubilization of metals from biosolids-treated soils. The experiment used two biosolids-treated Romona soils and the biosolids-free Romona soil control that was collected from the experimental plots of a long-term biosolids land application experiment (Hyun et al., 1998). The metal contents of the soils used were obtained by microwave-digesting 0.2 g soil with a mixture of 1.0 mL H_2O, 4.5 mL concentrated HNO_3, and 1.5 mL concentrated HCl, in a 120 mL Teflon PFA digestion vessel at 600 W full power with 100% power increment for 20 minutes and with maximum pressure of 484 kPa (Table 2-3).

Table 2-3. Chemical Properties of the Biosolids-treated Romona Soils.

Biosolids Treatment[a]	pH[b]	Total Soil Concentration (mg kg^{-1})					
		Cd	Cr	Cu	Ni	Pb	Zn
Control	7.7	0.5	37	105	24	25	95
135 Mg ha^{-1}	6.9	11	243	188	88	120	559
1,080 Mg ha^{-1}	6.1	26	596	478	215	396	1466

[a]Obtained at Moreno Field Station of the University of California, Riverside, CA. From 1976 through 1981 composted biosolids were applied at dry weight rates of 0 (control), 22.5 and 180 Mg ha-1 yr-1, respectively.
[b]1 : 1 w/v ratio

Metal dissolution experiments were conducted with 1 g of soil mixed with 10 mL of organic acids mixtures in 50 mL Teflon test tubes. The contents were allowed to equilibrate using a rotary mixer, which rotated the capped test tubes head to tail at approximately 1 rpm for 48 hours. One mL of chloroform was added to each test tube to prevent microbial decomposition of organic acids during the course of the equilibration. Three total organic acid concentrations 0.001, 0.01, and 0.1M in 13.5 mM $Ca(NO_3)_2$ along with a 13.5 mM $Ca(NO_3)_2$ blank were tested. The pHs of the organic acids mixtures were >4.2, 3.5–4, and 3–3.2 for 0.001, 0.01, and 0.1 M, respectively. Each treatment combination was replicated two times. After the equilibration, the soil suspensions were centrifuged to separate the solution and solid phases. The metal contents of the supernatants were determined using ICP-OE spectroscopy. The same experimental set up was used for the batch metal dissolution kinetics studies. The metal concentrations were determined at 0, 0.25, 0.5, 1, 2, 4, 8, 24, and 48 hours.

2.3 Field Validation

Two long-term biosolids land application sites were employed in the field validation of the above referenced metal phytoavailability mathematical model: (1) the biosolids land reclamation site of Metropolitan Water Reclamation District of Greater Chicago at Fulton County, Illinois and (2) the long term biosolids land application experiment at the Rosemount Experiment Station, University of Minnesota, St. Paul, Minnesota. The field data were provided by Dr. Thomas C. Granato of Metropolitan Water Reclamation District of Greater Chicago and Dr. Robert Dowdy of University of Minnesota, respectively.

Details on the nature of the data and procedures used in the field validation are described in Chapter 8.0.

CHAPTER 3.0

LITERATURE REVIEW – ORGANIC ACIDS IN ROOT EXUDATES

The chemical and biological reactions occurring in the soil-root interface play an important role in the availability of metals to plants. The soil factors include metal concentration in soil, interactions of metals with the soil surfaces, and the pH at the root-soil interface. The plant roots may change the physical, chemical, and biological conditions of the soil immediately adjacent to the root, referred to as the rhizosphere. In comparison to bulk soils, the rhizosphere is enriched with organic substances of plant and microbial origins that include organic acids, sugars, amino acids, lipids, coumarins, flavonoids, proteins, enzymes, aliphatics, aromatics, carbohydrates, and other substances (Schilling et al., 1998; Steele et al., 1999; Ali et al., 2000; Pomilio et al., 2000). Among them, the organic acids are the most significant as substrates for microbial metabolism and precursors of chemical reactions. The most commonly found organic acids in the rhizosphere are acetic, butyric, citric, fumaric, lactic, malic, malonic, oxalic, propionic, tartaric, and succinic acids. Organic acids are involved in chemical processes that release the not so readily available nutrients such as iron, phosphorus, and metals in soils for plant absorption.

Under the natural setting, trace metals such as Cd, Cr, Cu, Ni, Pb, and Zn frequently are present in the soil in low concentrations and tend to form sparingly soluble minerals or be adsorbed by other solid phases. Bioavailability of these metals is dependent on the biogeochemical interactions among soil, plant, and microorganisms in the rhizosphere. At the root-soil interface, substances produced by plants may affect the chemical reactions and biological activities in the soil and vice versa. Jenny and Overstreet (1938) postulated that plants absorbed mineral elements through contact and ion exchange between soil colloids and roots. Eaton (1936) reported that plants absorbed P and Fe from insoluble minerals across particle-root interfaces.

3.1 Root Exudates

Root exudates are substances in the rhizosphere soil solution that are secreted by roots of living plants and microbial modified products of these substances. They consist of low molecular weight organic compounds secreted by living and intact roots (Rovira, 1969; Rovira, 1979), freely and passively released root cell material, and polysaccharidic mucilage material associated with roots. Roots of all higher plants produce exudates that are complex mixtures of organic acids, reduced sugars, amino acids, phenolic compounds, and small amounts of other less easily identified substances (Table 3-1). Among them, organic acids are by far the most commonly found and the most abundant components of root exudates.

3.2 Factors Affecting Root Exudates

3.2.1 pH of Rhizosphere

The pH of the rhizosphere is important in determining metal and nutrient mobilization and uptake. It also affects microbial activity in the vicinity of the root. Root-induced pH change in the rhizosphere is a well-known phenomenon (Nye 1981; Marschner et al., 1986) and has an effect on the availability or solubility of nutrients such as P, Fe, Mn, Zn, Cu, and Al (Gahoonia, 1993). Rhizosphere pH may differ from the bulk soil pH by more than 2 units (Marschner et al., 1987). Buffer capacities of soil and root activity are the main factors influencing the pH at the soil-root interface (Youssef and Chino, 1989). Several hypotheses have been postulated to explain the differential ability of plant species in affecting rhizosphere pH: these include differences in root exudation and respiration patterns and differences in cation/anion uptake rates (Marschner and Romheld, 1983).

Table 3-1. Substances Found in Plant Root Exudates.

Category	Substances Identified
Amino acids and amides	α-Alanine, β-alanine, γ-amino butyric acid, arginine, asparagine, aspartic acid, cystathionine, cystine/cysteine, glutamic acid, glutamine, glycine, histidine, leucine/isoleucine, lysine, methionine, phenylalanine, proline, serine/homoserine, threonine, tryosine, tryptophane, valine
Enzymes	Amylase, invertase, phosphatase, protease, polygalacturonase, xylanase
Fatty acids	Linoleic, linolenic, oleic, palmitic, stearic
Growth factors	ρ-Amino benzoic acid, auxins, biotin, choline, inositol, η-methyl nicotinic acid, niacin, pantothenate, pyridoxine, thiamine
Nucleotides and flavonones	Adenine, cytidine, flavonone, guanine, uridine
Organic acids	Acetic, adipic, butyric, citric, fumaric, glutaric, glycolic, hydroxyglutaric, lactic, maleic, malic, malonic, oxalic, propionic, pyruvic, succinic, tartaric, valeric
Phenolic acids and coumarins	Caffeic acid, cinnamic acid, coumarin, o-coumaric acid, ρ-coumaric acid, ferulic acid, ρ-hydroxybenzoic acid, protocatechuic acid, salycilic acid, syringic acid, vanillic acid
Sterols	Campesterol, cholesterol, sitosterol, stigmasterol
Sugars	Arabinose, fructose, fucose, galactose, glucose, maltose, oligosaccharide, raffinose, rhamnose, ribose, sucrose, xylose
Others	Fluorescent substances, glycosides, glucosides, hydrocyanic acids, organic phosphorus compounds, polypeptide, saponin, sorgoleone, vitamins

Since plant roots acquire most mineral nutrients and metals as ions, imbalances between the absorption of cations and anions result in root excretion of compensating H^+ or OH^- ions into the soil in the rhizosphere, to prevent changes in the electroneutrality of the root tissues. (Darrah, 1993). Nitrogen can be supplied as either nitrate and/or ammonium and the form of nitrogen taken up can have an important effect on the rhizosphere pH (Marschner et al., 1982). The pH decreases when ammonium is taken up, and increases with nitrate uptake due to an unequal net uptake of cation and anion equivalents (Marschner et al., 1986). When ammonium is the major N source, the uptake of cations will exceed the uptake of anions. The surplus positive charges are believed to be balanced by the release of protons (Muranyi et al., 1994). Riley and Barber (1971) reported that fertilizing soybean (*Glycine max.* L.) with NH_4^+-N caused the rhizo-cylinder pH to decrease. The difference in rhizosphere pH between treatments was as large as 1.9 pH units (with an initial pH of 5.2) and as small as 0.2 units when the soil pH was 7.8 prior to the N application. However, very few detailed studies have measured rhizosphere pH for different plant species and genotypes (Romheld and Marschner, 1984). Studies of the rhizosphere changes in pH along roots of crops and the pH changes at different distance from roots are presented in Table 3-2. Gollany and Schumacher (1993) conducted a growth chamber study to characterize patterns of pH change within the rhizosphere of plants and the pH at different root zones (maturation, elongation, meristematic) was measured by a microelectrode at 1, 2, 3, and 4 mm from the root surface. They reported that the pH decreased to 4.82 and 4.95 in the rhizosphere around elongation and meristematic zones, respectively, compared to the control (pH of 7.6) without plants. Zhang and Pang (1999) demonstrated that while pH was essentially uniform in unvegetated soil, in the vegetated soils, the pH was lower near the root tips than at other locations around the root. The pH varied from 6.2 near the surface (0.5 cm depth) to 4.99 at 6 cm below the soil surface. The pH changes in the rhizosphere of selected genotypes of barley, *Hordeum vulgare* L. (Youssef and Chino, 1989; Gollany and Schumacher, 1993), corn (Fisher et al., 1989; Gollany and Schumacher, 1993), rape (Ruiz and Arvieu, 1990), and soybean (Romheld and Marschner, 1984; Youssef and Chino, 1989; Gollany and Schumacher, 1993) are given in Table 3-2.

Table 3-2. Reported pH Ranges of the Rhizosphere.

Plant	Genotype	pH Range	Reference
Barley	Bowman	6.2 - 7.6	Gollany and Schumacher (1993)
	Primus II	6.0 - 7.8	
	Dorirumugi	4.8 - 7.1	Youssef and Chino (1989)
Corn	Pioneer-3737	5.2 - 7.6	Gollany and Schumacher (1993)
	Pioneer-3732	5.2 - 7.6	
	CM-37	6.0 - 7.6	
	-	4.8 - 6.7	Fisher et al. (1989)
Clover	Trikkala	6.2 - 7.1	Hinsinger and Gilkes (1996)
Daisy fleabane	-	5.1 - 6.3	Zhang and Pang (1999)
Nectarine tree	Maxim	5.3 - 8.2	Tagliavini et al. (1995)
Oat	Hytest	6.0 - 7.6	Gollany and Schumacher (1993)
	SD 84104	6.2 - 7.6	
Canola (Rape)	-	5.7 - 6.4	Ruiz and Arvieu (1990)
Rye	Standard	5.6 - 7.1	Hinsinger and Gilkes (1996)
Sordan	S-757	6.0 - 7.6	Gollany and Schumacher (1993)
	S-333	6.1 - 7.6	
Sorghum	SC-33-8-9EY	6.3 - 7.6	Gollany and Schumacher (1993)
	SC-118-15E	6.6 - 7.6	
Soybean	Hawkeye	5.5 - 7.1	Romheld and Marschner (1984)
		4.8 - 7.6	Gollany and Schumacher (1993)
	PI-54169	5.8 - 7.6	Gollany and Schumacher (1993)
	Toyosuzu	5.1 - 7.0	Youssef and Chino (1989)
	-	4.7 - 7.1	Riley and Barber (1971)

3.2.2 Other Factors

The quantity and composition of root exudates are affected by environmental factors (Table 3-3). Uren and Reisenauer (1988) reported that acidification, chelation, precipitation, and redox reactions in the soil changed the nature and composition of root exudates. Nutritional stress experienced by plants, such as deficiencies of P, Zn, and Fe, could increase root exudation (Graham et al., 1981; Zhang et al., 1989; Kato-Noguchi et al., 1994; Romheld and Awad, 2000; Yang and Crowley, 2000; Rengel and Romheld, 2000). The presence of microorganisms in the growth media might also increase root exudation (Barber and Martin, 1976; Martin, 1977; Prikryl and Vancura, 1980; Schonwitz and Ziegler, 1982; Gardner et al., 1983c; Uren and Reisenauer, 1988; Meharg and Killham, 1991; Szmigielska et al., 1995; Burken and Schnoor, 1996). Under severe Zn deficiency conditions, root exudation of low molecular weight organic solutes, such as amino acids, sugars and phenolic organic compounds in both monocots (i.e., wheat) and dicots (i.e., cotton and soybean) were increased (Cakmak and Marschner, 1988b). Barber and Gunn (1974) and D'Arcy-Lameta (1986) showed that soybean grown on the solid growth media secreted greater amounts of root exudates than those cultured in the solution plant-growing medium. Temperature and photoperiod have profound effects on the quality and quantity of root exudates, because they affect the processes of photosynthesis, translocation, and respiration in plants (Hale et al., 1971; Hale and Moore, 1979). Increase in exudation at high temperature has been reported for many crops (Rovira, 1969; Hale et al., 1978; Uselman et al., 2000). Pramanik et al. (2000) demonstrated that the rate of root exudation of some organic acids for plants in vegetative and reproductive stages increased with the elevation of temperature and the elongation of photoperiod, and the mean rate was two or more times higher than the minimum exudation at low temperature with short photoperiod.

The imbalance in plant absorption of cations and anions affects the concentration of organic acids in the rhizosphere. In situations where roots take up an excess of cations, particularly K^+, negative charges that are provided by organic acids (such as malic, malonic, oxalic, and citric acids) are required to maintain an electrostatic balance (Osmond, 1976; Chang and Roberts, 1991). Roots of plants grown in soils receiving $NH4^+$-N have lower concentrations of the organic acids than those grown in soils receiving $NO3^-$-N (Marschner, 1995; Ali et al., 2000).

Table 3-3. Factors Affecting Root Exudates Production.

Factor	Effects	Reference
Soil pH	Acidification of the rhizosphere changes the composition of root exudates.	1
Soil texture	Roots of plants growing in sand produced greater amounts of exudates than those grown in clay or nutrient solution because of abrasive injury that is caused by the sand particles.	2
Reduction Potential	Plants produce root exudates of different composition under aerobic or anaerobic condition.	3
Light intensity	Light intensity affects rates of photosynthesis of plants that in turn affects nutrient translocation at the rhizosphere and the secretion of root exudates. High light intensity increases root exudates.	4
Soil temperature	Temperature affects rates of photosynthesis, translocation of photosynthates to the root, and enzymatic reactions. High temperature stimulates root exudation.	5
Soil moisture content	When a root system experiencing water stress is irrigated, it causes an increase in root exudation.	6
Nutrient availability	P, Fe, Zn, and Cu deficiencies increase excretion of root exudates.	7
Plant species and cultivars	Root exudates of different plants and cultivars have different chemical compositions.	8
Rhizosphere physical properties	Different physical structures in rhizosphere could form different microhabitats for different microorganisms and induce different root exudates. Cluster roots have been associated with localized increased concentration of root exudates.	9
Age of the root part	Exudates are secreted more frequently by roots during the initial stages than the later stages of growth. Mother roots exude more root exudates than mycorrhizal roots. Root exudates come primarily from the emerging lateral root tips.	10
Microorganism	Presence of microorganisms increases root exudation and they use exudates as growth substrates. Bacterial population lowers exudation of mycorrhizal roots and changes in exudate compositions.	11
Stress condition	Stresses caused by nutrient deficiency, phytotoxicity, water stress, O_2 and CO_2 concentration, temperature extremes, pH and salt concentration can change the composition of root exudates.	12
Cation-anion balance	Where roots take up an excess of cations, the negative charge required to balance this is often provided by organic acids.	13
NH_4^+/NO_3^- ratio	Roots grown under NH_4^+ have lower organic acid concentrations than those grown under NO_3^-.	14
Salinity	More organic and amino acids are produced under low salinity.	15

[1]Meharg and Killham, 1990; Misra and Ramani, 1992; Szmigielska et al., 1995; [2]Burken and Schnoor, 1996; Iijima et al., 2000; [3]Wiedenroth and Poskuta, 1981; [4]Hale et al., 1971; Nye et al., 1977; Hale and Moore, 1979; Graham et al., 1982; Misra and Ramani, 1992; Kato-Noguchi et al., 1994; Pramanik et al., 2000; [5]Rovira, 1969; Hale et al., 1979; Graham et al., 1982; Ali et al., 2000; Pramanik et al., 2000; [6]Nye et al., 1977; [7]Nye et al., 1977; Kraffczyk et al., 1984; Laheurte et al., 1990; Pojasok and Kay, 1990; Anderson, 1993; Kato-Noguchi et al., 1994; Szmigielska et al., 1995; Schilling et al., 1998; Marschner, 1998; Yang and Crowley, 2000; Rengel and Romheld, 2000; [8]Smith, 1970; Zhang et al., 1989; Laheurte et al., 1990; Anderson, 1993; Kato-Noguchi et al., 1994; Szmigielska et al., 1995; Bozhkov et al., 1996; [9]Chaboud and Rougier, 1991; [10]Schilling et al., 1998; [11]Barber and Martin, 1976; Martin, 1977; Prikryl and Vancura, 1980; Schonwitz and Ziegler, 1982; Gardner et al., 1983c; Uren and Reisenauer, 1988; Meharg and Killham, 1991; Szmigielska et al., 1995; Burken et al., 1996; Groleau-Renaud et al., 2000; [12]Graham et al., 1981; Zhang et al., 1989; Kato-Noguchi et al., 1994; [13]Osmond, 1976; Chang and Roberts, 1991; [14]Marschner, 1995; Ali et al., 2000; [15]Perez-Alfocea et al., 2000.

Root exudates affect the availability of nutrients by directly influencing the chemistry of the rhizosphere soils and indirectly enhancing microbial activity in the rhizosphere soils (Helal and Sauerbeck, 1984; Jones et al., 1994; Groleau-Renaud et al., 2000). Microorganisms utilize sugars, organic acids, amino acids, vitamins, and high molecular weight compounds exuded by roots as sources of carbon and substrates for energy (Lynch, 1982). Microbial metabolites may in turn increase root exudation through stimulation of root activity and growth, thus increasing the surface area over which the exudation occurs. Kozdroj and Van Elsas (2000) used PCR based on 16S rDNA sequence followed by denaturing gradient gel electrophoresis (PCR-DGGE) to evaluate the effects of organic acids in root exudates as a potential mechanism for selectively enhancing specific bacterial populations in contaminated soils, leading to the development of shifted communities differing in qualitative and quantitative composition (Yang and Crowley, 2000). The results indicated that root exudates significantly affected the development of bacterial populations in soil contaminated with heavy metals and, depending on the availability and conditions prevailing in the habitat, different bacterial populations were stimulated. When the microorganisms were active, the content of sugar and amino acids in root exudates decreased while organic acids increased (Graham et al., 1981; Dixon et al., 1989; Azaizeh et al., 1995). Waschutza et al. (1992) reported that, when bacteria were introduced into a sterile hydroponics culture, bacteria utilized glucose, malate, and succinate excreted by the plant roots as growth substrate and produced metabolites such as acetate, pyruvate, and ribose. As a result, the amounts of organic and amino acids in the nutrient solution remained unchanged, while sugar contents in the nutrient solution steadily decreased with the introduction of the bacteria.

Root exudation may mobilize mineral nutrients in the soil by acidifying the rhizosphere soils, thus solubilizing adsorbed and/or precipitated mineral nutrients. Plant roots physiologically have an ability to modify the ionic environment in the rhizosphere by exudation of protons and organic acids. These exudates affect the pH of the rhizosphere and the metal sorption and solubility processes. Enhanced acidity and organic anion concentrations in the rhizosphere can increase metal availability to plants under the stress of metal deficiencies. They may also keep nutrient elements in soil solution by chelating the dissolved metallic ions (Romheld and Marschner, 1986; Fan et al., 1997). Much of the work has been directed at examining interactions with Fe, Mn, Cu, and Zn (Takagi et al., 1988; Awad et al., 1988; Treeby et al., 1989; Zhang et al., 1991), and implications of organic acids on P dissolution in soils (Dinkelaker et al., 1989). In addition to their role in plant nutrition, root exudates may affect vital soil ecosystem processes such as the dynamics of soil organic matter transformation, cycling of nutrients in soils, and development of stable soil structure (Swinnen et al., 1994; Bozhkov et al., 1996; Traore et al., 2000). Czarnes et al. (2000) demonstrated that root exudates stabilized the soil structure in the rhizosphere by increasing the strength of bonds between particles.

3.3 Plant Species and Culturing Methods

3.3.1 Plant Species Investigated

Many plant species have been used in root exudation studies. Tested species have included:

- **field crops** [corn, wheat, barley, rice (*Oryza sativa* L.), sorghum (*Sorghum bicolar* L.), soybean, cotton (*Gossypium hirsutum* L.), alfalfa (*Medicago sativa* L.), rye (*Elymus dauricus* L.), and oat (*Avena sativa* L.)];

- **vegetable crops** [cucumber (*Cucumis sativus* L.), tomato (*Lycopersicon esculentum* L.), pigeonpea (*Cajanus cajan* L.), lettuce (*Lactuca sativa* L.), and pepper (*Capsicum annuum* L.)];
- **trees** [poplar (*Populus deltoids* L.), spruce (*Picea abies* L.), pine (*Pinus sylvestris* L.), beech (*Fagus sylvatica* L.), and apple (*Malis sylvestris* L.)]; and
- **miscellaneous categories** [pasture grasses (*Agropyron repens* L. and *Bromus inermis* L.), tobacco (*Nicotiana tabacum* L.), sunflower (*Helianthus annuus* L.), and lupine (*Lupinus albus* L.).

Greater than 60% of the 250 root exudate studies reviewed for this study were conducted with corn, wheat, barley, sorghum, or soybean as the experimental plant. Corn was by far the most frequently used plant specimen in root exudation studies (Chaboud and Rougier, 1991).

In some incidences, plants were selected for their unique attributes. For instance, pigeonpea and rice were frequently used to study root exudation under P-deficient conditions because their roots had high capacities to absorb P even from forms of P not readily available to plants, such as the Fe- or Al-bound P (Otani et al., 1996).

3.3.2 Culturing Methods

In many root exudate investigations, plants were cultured under a controlled laboratory environment where they were grown under relatively clean or sterile conditions. Nutrient solution-filled filter paper in Petri dishes, nutrient agar in glass tubes, and static solution culture in sealed containers were commonly used to grow the experimental plants. In this manner, the accumulated exudates would not be affected by microbial degradation. Collected exudates were representative of the root secretions for the duration of growth. This approach provided adequate control on the experimental conditions and allowed the collection of unaltered and uncontaminated root exudates for chemical analysis. When plants were grown in growth chambers with nutrient media, artifacts might be introduced. Comparable plants grown in soils might release two to four times more carbon into the rhizosphere than those grown in hydroponics (Trofymow et al.,1987). Mechanical impedance, anaerobiosis, nutrient deficiency, and the presence of microorganisms all increased the rhizodeposition of carbon by plants.

Because of the technical difficulties in maintaining a sterile growing condition and accommodating rapid growth of plants in limited space, the investigations were limited to root exudates from plants that had been grown for short duration (Schonwitz and Ziegler, 1982; Haller and Stolp, 1985; Morel et al., 1986; Steele et al., 1999; Romheld and Awad, 2000; Pramanik et al., 2000; Groleau-Renaud et al., 2000). Approximately 50% of these studies examined the root exudates of plants that were two to four weeks old. However, the volume and chemical nature of root exudes are expected to vary with plant development (Smith, 1970; Nye et al., 1977; Hale and Moore, 1979; Zhang et al., 1989; Laheurte et al., 1990; Kato-Noguchi et al., 1994; Meharg and Killham, 1995; Bozhkov et al., 1996; Uselman et al., 2000).

3.3.3 Sterile Culturing Methods

Root exudates may be obtained from plants grown in sterile and non-sterile plant culture (Godo and Reisenauer, 1980; Morel et al., 1987; Awad et al., 1988; Mench et al., 1988a, 1988b; Takagi et al., 1988; Treeby et al., 1989; Zhang et al., 1989; Schilling et al., 1998). Early investigations almost exclusively employed aseptic plant culturing conditions. However, sterile conditions were difficult to maintain and plants were kept to a rather small size. More importantly, the sterile culturing conditions do not represent the realistic situations where plants are normally grown because they ignore the role of microbes in the formation of root exudates.

Most commonly, sterilized seeds were germinated in and allowed to grow in Petri dishes for a duration on absorption pads soaked with nutrient solution. Afterwards, the root exudates were extracted from the absorption pads for analyses (Mathews et al., 1989; Chaboud and Rougier, 1991; Kato-Noguchi et al., 1994; Szmigielska et al., 1995; Hodge et al., 1996; Nardi et al., 1997; Steele et al., 1999).

The seedlings germinated from sterilized seeds might be aseptically transplanted and grown in autoclaved glass tubes or sterile nutrient solution culture in growth chambers under phytotronic conditions (Morel et al., 1987; Mench et al., 1988a, 1988b; Mathews et al., 1989; Mench and Martin, 1991; Waschutza et al., 1992; Dakora et al., 1993; Szmigielska et al., 1995; Saeki et al., 1996; Kobayashi et al., 1996; Nardi et al., 1997; Pramanik et al., 2000). Antibiotic substances sometimes were used to maintain an aseptic rhizosphere environment (Otani et al., 1996; Fan et al., 1997; Subbarao et al., 1997; Siciliano et al., 1998).

3.3.4 Non-sterile Culturing Methods

Generally, seeds were sown in perlite or vermiculite and then transferred to hydroponics culture tanks with nutrient solution and grown for a given period of time (Schurr and Schulze, 1995; Otani et al., 1996; Burken et al., 1996; Subbarao et al., 1997; Fan et al., 1997; Pramanik et al., 2000; Groleau-Renaud et al., 2000). For examples, Zhang et al. (1989,1991) and Awad et al. (1994) germinated wheat and barley seeds in quartz sand with $CaSO_4$ solution in a growth chamber under non-sterile conditions and transferred seedlings to a continuously aerated nutrient solution. Seeds were germinated on moist filter papers in Petri dishes and transferred to hydroponics solution tanks (Friebe et al., 1995; Bozhkov et al., 1996; Yu and Matsui, 1997).

When cultured under septic conditions, plants were permitted to reach maturity (Habib et al., 1990; Soejima et al., 1992; Muller et al., 1993; Uselman et al., 2000). In these cases, investigators were interested in specific substances in the rhizosphere, such as polysaccharides, instead of the general chemical composition of root exudates.

3.4 Methods for Root Exudates Collection

Following the cultivation, root exudates must be separated from the growing media. If plants are grown in solution culture, aliquots of the nutrient solution can be purified and concentrated for the assays. The plants also can be transferred from nutrient solution to another receptacle where root exudates are deposited in water. If plants are grown in solid media, the root exudates are first flushed out by water. The root exudates in the solution then are separated and concentrated for analysis.

Both septic and aseptic approaches have been used for collecting root exudates. In one approach using aseptic procedures, plants were removed from the growing media, rinsed to clean the roots, and the root treated with antibiotic agents. The antibiotic-treated plants then were allowed to accumulate the root exudates in an aseptic environment for two to six hours. Li et al. (1997) reported that the acid phosphatase activity of the rhizosphere in the non-sterile treatment was not significantly different from that in the sterile treatment.

3.4.1 Collection under Sterile Conditions

Root exudates of plants cultured under aseptic conditions must be handled in sterile environment prior to the analysis (Chaboud, 1983; Mench et al., 1988a, 1988b; Meharg and

Killham, 1991; Mench and Martin, 1991; Barbour et al., 1991; Szmigielska et al., 1995; Meharg and Killham, 1995; Hodge et al., 1996). At the end of the growing period, the solution in nutrient tubes or hydroponics solutions were collected and filtered by 0.2 or 0.45 μm membrane filters. The sterile filtrates that contained soluble root exudates were concentrated by rotary evaporation (Saeki et al., 1996; Nardi et al., 1997) or by lyophilization. The collected material was stored at –20° to 4°C (Chaboud and Rougier, 1991; Pant et al., 1994) and dissolved in a known volume for analysis (Mathews et al., 1989; Fan et al., 1997).

Some researchers employed anti-microbial agents during the root exudate collection process for plants cultured under non-sterile conditions. In this process, the roots were removed from the growth media and treated with an antibiotic solution such as chloramphenicol and chloroform (Otani et al., 1996), chloramphenicol and thymol (Subbarao et al., 1997), or Micropur (a commercially available silver-based anti-microbial agent) (Fan et al., 1997; Gransee and Wittenmayer, 2000) and then rinsed with sterile distilled water. After the antibiotic treatment, the roots were soaked under aseptic conditions in distilled water to accumulate root exudates for a period of time. The root exudates in the soaking solution were concentrated using a rotary evaporator.

3.4.2 Collection under Non-sterile Conditions

Exudates from intact plant roots are collected, by placing the roots in aerated double-distilled water for two to six hours. The exudate solution is then filtered and concentrated at 50°C under rotary evaporation to 20 mL for analysis of the organic solutes (Zhang et al., 1989,1991).

Tang and Young (1982) and Muller et al. (1993) modified the technique used for collecting allelopathic substances. They demonstrated that root exudates might be adsorbed on XAD-4, a porous polymer resin. Root exudates trapped on the resin were extracted either with distilled water (Friebe et al., 1995; Yu and Matsui, 1997), methyl alcohol (Masaoka et al., 1993), or simply tap water for lipids (Bozhkov et al., 1996). Visser et al. (1987) applied this technique successfully in the recovery of germination stimulants for the parasitic angiospermous seeds. The system employed by Visser et al. (1987) has several advantages over previously used extraction procedures and is easily adaptable to large-scale recovery and extremely low-level concentrations of substances, as is very often the case with germination stimulants and other biologically active exudates.

3.5 Analysis Methods for Root Exudates

3.5.1 HPLC and GC-MS

Only a few attempts have been made to comprehensively analyze the chemical composition of root exudates. This is due to the necessity of employing a wide array of analytical methods for characterizing the different exudate components. Prior knowledge on the chemical nature of root exudates substances is essential for selecting the methodologies for proper separation, identification, and quantification of individual compounds from the mixtures. Concentrations of substances in root exudates may be low, further complicating their detection and quantification.

The chemical behavior of low molecular weight organic acids commonly found in root exudates are similar, making their separation in a chromatographic column a challenging analytical problem. Most high performance liquid chromatographic (HPLC) columns lack specificity and could not completely separate a mixture of mono-, poly-, and hydroxycarboxylic acids in a single chromatographic run (Blake et al., 1987). Szmigielska et al. (1995) experimented with a gas chromatographic (GC) approach based on the differences in the chemical structure of the organic acids.

Several HPLC procedures were employed in root exudate assays. They included:

- Ion exclusion using a strong cation exchange resin to remove paramagnetic ions (Zhang et al., 1989 and 1991; Foy et al., 1990; Petersen and Bottger, 1991; McFeeters, 1993; Engels and Marschner, 1993; Fan et al., 1997; Young et al., 1998);
- Ion exchange using a weak anion exchange resin (Mench and Martin, 1991; Lilieholm et al., 1992; Waschutza et al., 1992; Nardi et al., 1997; Li et al., 1997; Gransee and Wittenmayer, 2000); and
- Reversed phase extraction with or without sample derivatization (Miwa, 1985; Muller et al., 1993; Kato-Noguchi et al., 1994; Morse et al., 2000).

Depending on the substances to be assayed, three HPLC systems were used:

- Sugars: CarboPac TM PA 1 (Dionex) with Detector PAD/Gold/RT;
- Organic Acids: Aminex HPX-87H with Detector UV 210 nm; and
- Amino Acids: Spherisorb ODS2/RT with Detector fluorescence Ex 330 nm and Em 450 nm (Waschutza et al., 1992).

Gas chromatography with mass spectrometry (GC-MS) analysis was used to quantify the changes of quantities of root exudate in relation to colonization of roots by mycorrhizal fungi and the concentrated root exudate solution was injected directly into the instrument (Marschner et al., 1997). Nardi et al. (2000) used GC-MS measurements to study the different soil organic matter mobilization by agrarian and forest root exudates. To enhance the resolution, root exudates might be separated into cationic, anionic, and neutral fractions using ion exchange resins prior to the analysis on a GC-MS (Otani et al., 1996). Compounds were identified by the retention time and mass spectra of respective reference compounds (Mench et al., 1988a, 1988b; Pojasok and Kay, 1990; Szmigielska et al., 1995; Friebe et al., 1995; Pramanik et al., 2000).

3.5.2 Other Instrumentation Approaches

The chemical composition of root exudates can be determined by a combination of several methods. For examples, chromatographic methods are used first to separate root exudates into simpler chemical components. The well resolved components are collected and purified for structural analysis using nuclear magnetic resonance (NMR) methods (such as ^{1}H-NMR, ^{13}C-NMR, and multi-dimensional NMR), mass spectroscopy (MS), infrared spectroscopy (IR), or ultraviolet spectroscopy (UV) (Soejima et al., 1992; Dakora et al., 1993; Kobayashi et al., 1996; Nardi et al., 1997; Fan et al., 1997; Schilling et al., 1998). Steele et al. (1999) used capillary zone electrophoresis coupled to a UV spectral array detector (CZE-UV), high performance thin-layer chromatography with densitometry (HPTLC-UV), and GC-MS to analyze the root exudates. Pomilio et al. (2000) separated and purified the constituents of the root exudates of wild oats using HPTLC, capillary GC, and medium-pressure liquid chromatography (MPLC) and characterized the compounds by GC-MS and ^{1}H-NMR. Fan et al. (1997) used the ^{1}H and ^{13}C 2-D NMR techniques to analyze root exudates of barley grown on hydroponics culture. By coupling the NMR data with the GC-MS data of the corresponding samples, they obtained complete structural information on organic and amino acids and several previously unidentified exudate components. These procedures provided separation, identification, and structural confirmation of the previously unidentified root exudates and are appropriate for examining a variety of environmental effects in a short time period. These methods, however, were not intended for routine analysis. It nevertheless illustrated the possibility that a large number of chemical components in the root exudates could be identified.

Other methods used to analyze specific components in root exudates included: AAS for cations, HPLC for amino acids, ion-exchange chromatography for anions, and ELISA, a Bio-Kinetics Microplate Reader, for abscisic acids (Schurr and Schulze, 1995), GC for mono-saccharides and amino acids, HPLC for free organic acids, and Carbon-Hydrogen-Nitrogen elemental analyzer for total carbon (Laheurte et al., 1990; Mench and Martin, 1991; Masaoka et al., 1993).

3.6 Composition and Concentration of Root Exudates

Plant roots secrete a complex mixture of soluble exudates which may contain sugars, amino acids, organic acids, lipids, coumarins, flavonoids, proteins, enzymes, aliphatics, aromatics, low molecular weight carbohydrates, and other substances (Rovira, 1969; Vancura and Hanzlikova, 1972; Rovira and Davey, 1974; Hale and Moore, 1979; Chhonkar and Tarafdar, 1981; Mench et al., 1988a, 1988b; Vaughan and Ord, 1991; Awad et al., 1994; Pant et al., 1994; Vempati et al., 1995; Hodge et al., 1996; Mergel et al., 1996; Marschner et al., 1997; Fan et al., 1997; Schilling et al., 1998; Steele et al., 1999; Ali et al., 2000; Pomilio et al., 2000; Gransee and Wittenmayer, 2000; Wittenmayer and Szabo, 2000). The rhizosphere soil solution also contains cell lysates, which are released when root cortex cells are ruptured either by mechanical damage, grazing, or the emergence of lateral roots that break through the root epidermis. In addition to these substances, rhizodeposition includes the secretion of enzymes, such as acid phosphatase secreted in response to P deficiency (Helal and Dressler, 1989; Tadano and Sakai, 1991). Most investigations, however, focused on one specific group or family of related compounds in the root exudates.

Although the identified components differed in molecular weight, amount released, and biochemical role, the dominant exudate components were always shown to be organic acids, sugars, and amino acids. Smith (1970) and Barber and Gunn (1974) made quantitative measurements of amino acids and organic acids exuded by plant roots. Models of organic acid accumulation in the rhizosphere suggested that concentrations as high as 50 mM could occur within 1 mm from the root surface (Jones et al., 1996b). Typically, the total organic acid concentration in roots is around 10 to 20 mM (1-4% of total dry weight), which is comparable, at least for corn, with the other main organic solutes present in root cells, namely amino acids (10 to 20 mM) and sugars (90 mM) (Jones and Darrah, 1994, 1996). The major organic acids of plant-origin have been detected in the rhizosphere in appreciable quantities (0.1 to 100 mM) and constituted about 2% of the total dissolved organic carbon in podzolic soils (Kryzszowska et al., 1996; Shen et al., 1996). Generally higher concentrations are found in rhizosphere soil compared to those present in the bulk soil. The carboxylic acids that played a significant role in complexing metal ions in soils were not utilized as rapidly as carbohydrates by the soil microorganisms. Kraffczyk et al. (1984) found that root exudates released under aseptic conditions contained higher concentration of monosaccharides than those released under septic conditions did, while the concentrations of organic acids and amino acids were similar under both sterile and non-sterile treatments. It appeared that microorganisms preferentially utilized the carbohydrates in root exudates (Jones and Darrah, 1994; Jones et al., 1996a). As a result, the organic and amino acids became the prevalent chemical substances in root exudates (Inskeep and Comfort, 1986).

3.6.1 Organic Acids

Typical organic acids found in root exudates are malic, malonic, acetic, citric, fumaric, succinic, lactic, tartaric, oxalic, and others. Organic acids of root exudates seem to have a major role in solubilization of mineral nutrients. Malic and citric acids are major root exudate components for some plant species under aseptic condition. For example, in corn grown under sterile conditions, exudation of malic and citric acids increases significantly under nutrient deficiency (Jones and Darrah, 1993ab). Piscidic acid, a strong chelater of metal ions found in the root exudates of pigeonpea, may be involved in the chelation of Fe from sparingly soluble iron phosphates, thus making P in the soil available to the plant (Ae et al., 1990a, 1990b, 1993; Subbarao et al., 1997; Schilling et al., 1998). Some researchers reported that citric acid was the most abundant organic acid in the rhizosphere where lupine was cultured under sterile conditions (Gardner et al., 1983a, 1983b; Kraffczyk et al., 1984; Li et al. 1997). In addition, small quantities of malic acid and oxalic acid were also detected. The relative abundance of the low molecular weight organic acids identified in root exudates of several wheat cultivars grown under the sterile hydroponics systems followed the order: succinic > oxalic > malonic > fumaric (Engels and Marschner, 1993; Ochs et al., 1993; Cieslinski et al., 1994; Szmigielska et al., 1995). Malonic acid was a major component for some plants such as pigeonpea (Otani et al., 1996; Nardi et al., 1997; Young et al., 1998). Delhaize et al. (1993) suggested that malic acid exuded by roots of wheat plants might be related to their tolerance or sensitivity to aluminum. Laheurte et al. (1990) found citric, malic, fumaric, gluconic, lactic, and succinic acids in the rhizosphere of pine and beech. Their relative abundance in the rhizosphere were fumaric > citric > malic and malonic > lactic > gluconic, for beech and pine, respectively. Organic acids also were present in root exudates of other crops. For instance, citric and/or malic acids were the major components of the root exudates of alfalfa (Lipton et al., 1987) and canola (Hoffland et al., 1989).

Under sterile conditions, many studies found that citric acid was one of the prevalent organic acids in root exudates (Dinkelaker et al., 1989; Laheurte et al., 1990; Li et al. 1997). Other researchers, however, demonstrated that citric acid was not detectable in the root exudates collected from plants cultured under non-sterile conditions (Zhang et al., 1991; Subbarao et al., 1997). Studies on the mineralization of exudates have shown that organic acids such as citric and maleic acids added at realistic rhizosphere concentrations are rapidly degraded in rhizosphere and bulk soil or may be converted to other organic acids in metabolic pathways (Krebs cycle) under septic conditions (Jones and Darrah, 1994; Jones et al., 1996a). The production of these organic acids is related to some biosynthetic pathways (Marschner, 1995). While citric and malic acid are involved in energy production as intermediates in the tricarboxylic acid (TCA) cycle, oxalic, malonic, and malic are primarily present in the cell for cation charge balancing or for maintaining osmotic potential.

3.6.2 Amino Acids

Compared to carboxylic organic acids, amino acids appear to play a less significant role in complexing or mobilizing plant nutrients. The proteinaceous amino acids are not readily detected in root exudate solutions. The lack of reaction between the amino acids and metals in solution may have reflected the short extraction times used in collecting root exudates (Jones et al., 1994).

The non-proteinaceous amino acids in the rhizosphere (phytosiderophores) have been extensively studied because they readily mobilize plant micronutrients in the soil. Marschner et al. (1986) investigated the release of phytosiderophores by roots of various Fe-deficient pasturegrasses, such as *Agropyron repens* L. and *Bromus inermis* L. species. This release follows a distinct diurnal pattern (Marschner et al., 1986; Nishizawa and Mori, 1987) and is particularly intensive in apical root zones (Mori et al., 1987). Mugineic acid, which has a low molecular weight, was the dominant non-proteinaceous amino acid (Romheld and Marschner, 1990). Under Zn and Fe deficiencies in growth media, wheat and barley roots released the phytosiderophore 2'-deoxymugineic acid and epi-3-hydroxy-mugineic acid, respectively (Zhang et al., 1989).

3.6.3 Aromatic Acids, Lipids, and Others

The role of aromatic acids on ion uptake by plant roots has been examined (Harper and Balke, 1981; Balke, 1985; Lyu and Blum, 1990; Booker et al., 1992). Phenols exuded by young soybean root hairs located on the primary and lateral roots play a significant role in the reduction of dissolved Fe^{3+} (Olsen and Brown, 1980; Olsen et al., 1981; Romheld and Marschner, 1983; Hether et al., 1984; Bell et al., 1988; Morse et al., 2000). The oxidation of phenols to diquinones released electrons and protons. Yu and Matsui (1994, 1997) reported that cinnamic acid was one of the root exudate constituents of cucumber.

Root exudates contained lipids such as free fatty acids, cholesterol esters, phosphatidyl choline, phosphatidyl ethanolamine, phosphatidyl inositol, and cardiolipin (Mathews et al., 1989; Bozhkov et al., 1996; Pramanik et al., 2000; Pomilio et al., 2000). Plant roots also released flavones or iso-flavones (Firmin et al., 1986; Horvath et al., 1986; Bassam et al., 1988; Steele et al., 1999). Dakora et al. (1993) demonstrated that alfalfa inoculated with *Rhizobium* bacteria could exude a phytoalexin (medicarpin). In the root exudates of rice and pine plants, several cytokinins also were identified (Gogala, 1989; Soejima et al., 1992; Ali et al., 2000).

3.7 Root Exudates and Plant Nutrient Interactions

Root exudates have a role in the translocation of trace elements in soils (Rovira et al., 1983; Curl and Truelove, 1986a, 1986b; Marschner et al., 1986; Mergel et al., 1992). They are directly related to acidification, metal chelation, precipitation, and oxidation-reduction reactions in the rhizosphere. Thus they influence the solubility and plant uptake of micronutrients, the dynamics of microbial activity, rhizosphere physical properties, and root growth patterns. There is increasing evidence that soluble root exudates increase the solubility of metals in the rhizosphere. Linehan et al. (1989) reported that concentrations of Cu, Zn, Mn, and Co in soil solution increased when plants were present. Morel et al. (1987) and Mench et al. (1988a, 1988b) showed that corn exudates, for example, were capable of binding Cd in soils. Knight et al. (1997) demonstrated that the hyperaccumulator *Thlaspi caerulescens* was highly efficient at mobilizing Zn that was not soluble initially. The role of root exudates in the mobilization of Cd and Zn in soils by the hyperaccumulators needs to be investigated.

3.7.1 Metal Complexation

The presence of organic acids in the rhizosphere influences the solution phase concentration of metals in the rhizosphere and, thereby, metals' availability to plants. Organic acids exuded by roots, such as oxalic, lactic, succinic and others, have been proven to be effective in releasing metals from sparingly soluble mineral forms by forming stable chelate complexes with the metal ions (Marschner, 1995). The level of complexation depends on the organic acid involved (number and proximity of carboxylic groups), the concentration and type of metal and the pH of the soil. Organic acids with only one carboxyl group, such as acetic, formic, and lactic acids, have very little metal complexing ability. Based on the stability constants summarized in Table 3-4, malic and oxalic acids have a high affinity for metals. These are the most frequently found organic acids in soils (Pohlman and McColl, 1986; Jones and Kochian, 1996). However, presence of these acids does not always entail efficient complexation at all soil pHs (Mench and Martin, 1991). For instance, using Geochem-PC (Parker et al., 1995), it can be illustrated that the complexation of Fe by malic and oxalic acids is highly dependent upon soil pH, with little complexation at high soil pHs. In addition, malic acid and oxalic acid have the tendency to precipitate in the presence of Ca^{2+}. This will reduce its potential complexing ability with some metals. In conclusion, plant roots can change metal concentration in the rhizosphere because of exudation of organic acids from the root to the rhizosphere.

In general, organic acids are able to induce a two to fourfold increase in mineral dissolution rate in comparison with rainwater alone. However, the extent of the chemical reactions is dependent upon mineral type, pH, and organic acid type (Pohlman and McColl, 1986; Jones and Kochian, 1996). The presence of organic acids increases the dissolution of Fe and Al oxyhydroxides (Jones and Kochian, 1996). More recently, adsorption phenomenon of organic acids on the hydroxyaluminum-montmorillonite (HyA-Mt) complex has been reported (Cambier and Sposito, 1991; Janssen et al., 1997). The added citrate did not appear to be adsorbed on the Al-OH groups of the HyA-Mt complex but only to the edge of the clay (Janssen et al., 1997). In contrast, Cambier and Sposito (1991) considered that the HyA-Mt complex was stable at $4 < pH < 5.5$, and only external HyA polycations could react with citrate. Sakurai and Huang (1995) studied the effect of oxalate on the adsorption of Cd by montmorillonite (Mt) and HyA-Mt complex at pH 5. The reaction was very rapid and virtually complete within 10 minutes. The presence of oxalate markedly interfered with the Cd adsorption reactions on the clays, especially on montmorillonite. Taniguchi et al. (2000) investigated the adsorption phenomena of Cd on the hydroxyaluminosilicate (HAS)-montmorillonite (Mt) and HyA-Mt complexes as influenced by oxalate and citrate. They concluded that the optimal concentrations of oxalate and citrate for Cd adsorption depended on the form of Cd ions and their proportions in the solution.

Table 3-4. Stability Constants of Organic Acids-Metal Complexes.

Organic Acid	M:L:H Ratio[a]	pK						
		H^+	Cd^{2+}	Cr^{2+}	Cu^{2+}	Ni^{2+}	Pb^{2+}	Zn^{2+}
Acetic Acid	1:1:0	4.76[b]	1.93[b]	1.25[c]	2.22[b]	1.43[b]	2.68[b]	1.57[b]
	1:2:0	—	3.15[b]	2.15[c]	3.63[b]	—	4.08[b]	1.90[d]
Butyric Acid	1:1:0	4.82[b]	1.25[e]	—	2.14[b]	2.14[b]	2.13[e]	0.98[e]
	1:2:0	—	1.98[e]	—	2.98[e]	2.98[e]	3.74[e]	1.65[e]
Glutaric Acid	1:1:0	5.43[b]	1.60[f]	—	3.16[b]	1.60[d]	2.48[f]	2.45[b]
	1:2:0	—	—	—	—	—	3.45[f]	1.74[g]
Lactic Acid	1:1:0	3.86[b]	1.70[b]	—	3.02[b]	2.22[b]	2.78[b]	2.22[b]
	1:2:0	—	2.10[g]	—	4.84[b]	2.76[g]	2.88[g]	3.75[b]
Maleic Acid	1:1:0	6.33[b]	2.40[d]	—	3.40[d]	2.00[d]	2.75[g]	2.00[d]
	1:2:0	—	3.60[h]	—	4.90[i]	—	4.03[g]	—
Oxalic Acid	1:1:0	4.27[b]	3.89[j]	3.85[d]	6.23[b]	5.16[b]	4.91[b]	4.87[b]
	1:2:0	—	—	6.81[d]	10.3[b]	—	6.76[b]	7.65[b]
Propionic Acid	1:1:0	4.87[b]	1.19[g]	—	2.22[b]	0.78[g]	2.08[e]	0.85[g]
	1:2:0	—	1.86[g]	—	3.00[k]	1.26[e]	4.15[b]	1.23[g]
Pyruvic Acid	1:1:0	2.55[b]	—	—	2.20[b]	1.12[l]	2.04[k]	1.26[l]
	1:2:0	—	—	—	4.90[b]	0.46[l]	3.40[k]	1.98[l]
Succinic Acid	1:1:0	5.64[b]	2.72[b]	—	3.28[b]	1.60[d]	5.64[b]	2.52[b]
	1:2:0	—	—	—	—	—	3.73[f]	2.00[g]
Tartaric Acid	1:1:0	4.37[b]	—	—	2.90[g]	2.06[g]	2.60[g]	3.82[b]
	1:2:0	N.D.	—	—	—	—	4.00[g]	5.00[b]
Valeric Acid	1:1:0	4.78[b]	—	—	2.08[b]	—	—	—

Source: Martell and Smith, 1976-1989; Stumm and Morgan, 1996

[a]M:H:L ratio indicates the Stoichiometry of the complex where M indicates metal, L indicates ligand, and H indicates proton.

[b]25°C, zero ionic strength in mol liter^{-1}; [c]25°C, 0.3 ionic strength; [d]25°C, 0.1 ionic strength; [e]25°C, 2.0 ionic strength; [f]30°C, 2.0 ionic strength; [g]25°C, 1.0 ionic strength; [h]25°C, 0.2 ionic strength; [i]18°C, zero ionic strength; [j]25°C, 0.16 ionic strength; [k]25°C, 3.0 ionic strength; [l]25°C, 0.5 ionic strength

N.D., not determined

3.7.2 Solubilization of Phosphorus

Phosphate dissolution rates can be greatly accelerated in soil in the presence of organic acids depending on soil type and speciation and concentration of organic acid (Gerke, 1994; Jones and Darrah, 1994). There are several possible mechanisms by which P release can occur. The first involves competitive sorption reactions (direct ligand exchange) in which exuded organic acids compete with P for common sorption sites on soil particles, thus releasing P from the solid phases. For example, citrate is able to desorb phosphate from sesquioxide surfaces by anion exchange (Parfitt, 1978). The second could involve the modification of the surface characteristics in which organic ligand sorption modifies the surface charge of P sorbents. The organic acids in the rhizosphere may be sorbed by the clay minerals, thereby increasing the net negative charge of clays (Chairidchai and Ritchie, 1992). Along with this, acidification due to proton exudation accelerates the dissolution reaction affecting clay minerals. The third involves the complexation of metal ions in the solid that constitutes the exchange matrix holding the P. The action of organic chelates in solubilizing phosphates and phosphate minerals is attributed to the formation of complexes with Ca, Fe, or Al, thereby releasing the phosphate in water-soluble forms (Stevenson, 1986).

Solubilization of P compounds by organic acids is accomplished by the formation of complexes of organic acid ligands with P-associated metal ions such as Fe, Al, and Ca. The organic ligands substitute for orthophosphates thus releasing P for plant uptake. This process depends on the number and position of carboxyl (-COOH) and phenolic (-OH) functional groups in the organic acids (Bolan et al., 1994; Li et al., 1997). The production of root exudates by crops grown under P-deficient or Al-toxic conditions was investigated (Hoffland et al., 1989; Ae et al., 1990a, 1990b; Otani et al., 1996). Mullette et al. (1974) indicated that citric and oxalic acids in root exudates of *Eucalyptus gummifera* chelated Al^{3+} and/or Fe^{3+} and released P in infertile soils. For canola, P might become available for plants, due to the presence of piscidic acid or other organic acids in the root exudates. The alcoholic, phenolic and carboxylic groups in the tartaric portion of piscidic acid were involved in the chelation of Fe of iron phosphates, thus, releasing P for uptake by the root (Subbarao et al., 1997). Studies with other species indicated that plants grown under P-deficient conditions produced several times more organic acids than comparable plants grown under normal conditions (Gardner et al., 1983a, 1983b; Young et al., 1998).

Mineralization of organic P and immobilization of orthophosphates are two major processes controlling the availability of soil P to plants. The transformation of organically bound P to orthophosphates is an essential process in the biochemical cycling of this element. It is facilitated by phosphatases whose activities were frequently enhanced in the rhizosphere (Halm et al., 1972; Tate, 1985; McLaughlan and Alston, 1986; Pant et al., 1994). Schilling et al. (1998) reported that root exudates affect the binding of phosphorus in soil and the P transport to the root. Root exudates played an important role in the utilization of insoluble phosphorus sources in a number of crop species. The solubilization of P minerals in soils was attributed to the presence of low molecular weight organic acids (Traina et al., 1986; Fox et al., 1990).

3.7.3 Solubilization of Fe

Organic acids in root exudates significantly affect amounts of total soluble Fe^{3+} in the rhizosphere. For example, the soluble Fe^{3+} in the rhizosphere solutions of *Pinus radiata* at pH 4 to 8.2 formed complexes with oxalate and aspartate in the root exudates (Inskeep and Comfort, 1986). Soluble organic ligands such as fulvic acid were thought to complex Fe^{3+}/Fe^{2+} in soil solution (Chaney, 1984). Hydroxamate siderophores have also been investigated for their contributions to soluble Fe in soil solution (Powell et al., 1982; Akers, 1983). Aggarwal et al. (1999) reported that galega tree (*Tephrosia purpurea* L.) released a compound that dissolved rock-iron [$Fe(OH)_3$]. This compound not only dissolved $Fe(OH)_3$ minerals but also had affinity for other trace metal ions. The primary mechanism for transport of Fe from soil solution to plant is diffusion. Outcomes of nutrient uptake simulation models indicate that root exudates may play an important role in the Fe transport processes (Nye et al., 1984). However, it is difficult to obtain the realistic data needed to accurately quantify root exudates, root morphology and microbial activities in the rhizosphere. Most studies of plant nutrition and physiology of Fe are conducted using Hoagland or Steinberg nutrient solutions containing Fe salts or Fe chelates.

These solutions, however, do not truly represent dominant labile Fe sources in soils (Vempati and Loeppert, 1988). In a natural soil environment, plants obtained Fe from Fe(III) amorphous oxides and/or other Fe-bearing clay minerals (Loeppert and Hallmark, 1985; Loeppert, 1985; Vempati and Loeppert, 1988; Loeppert et al., 1988).

Dicot plants are known to possess the capability to dissolve Fe enzymatically on the root surface (Romheld and Marschner, 1986). Fe-deficient growth media promotes the release of root exudates that may influence iron mobility (Porter, 1986; Mozarfar, 1989). Fe-deficient tomato plants exuded caffeic acid from their roots as a reductant (Olsen et al., 1981). The ability of caffeic acid to reduce Fe^{3+} to Fe^{2+}, however, was not as strong as the enzymatic reduction of Fe at the root surface of dicots (Romheld and Marschner, 1986). Soybean roots in particular also released citric acid during Fe-stress conditions (Hether et al., 1984; Marschner et al., 1986).

3.7.4 Solubilization of Zn and Other Metals

Components of root exudates such as citric, oxalic, tartaric, and acetic acids are capable of forming complexes with Cu, Zn, Pb, and Co ions in solution (Rovira, 1969; Merckx et al., 1986; Inskeep and Comfort, 1986; Mench et al., 1988a; Cakmak and Marschner, 1988a, 1988b) and enhance the mobilization and uptake of toxic heavy metals from contaminated soils (Treeby et al., 1989; Awad and Romheld, 2000). The role of root exudates in enhancing the uptake of Fe and toxic heavy metals from contaminated soils was investigated (Romheld and Awad, 2000). The results indicated that plants precultured in media with low Fe nutritional status had a higher uptake of Fe, Zn, Ni, and Cd (up to 200%) than control plants that were precultured in media with adequate Fe nutritional status. Various weak organic acids and/or their salts were tested for the remediation of a loam and a sandy clay loam soil naturally polluted by heavy metals for over three years against two strong synthetic chelating agents (EDTA and DTPA) by Wasay et al. (1998). Organic acids can be an attractive extracting agent because the extraction can be performed at mildly acidic conditions (pH 3-5) and they are biodegradable (Veeken and Hamelers, 1999).

Experimental evidence has shown that organic acids added to humic macromolecules induce the release of adsorbed metal ions (Nardi et al., 1986; Piccolo et al., 1992; Nardi et al., 1997). In near-neutral and neutral aqueous solution, organic acids are readily dissociated. The negatively charged organic acid ligands are capable of forming complexes with metals (such as Mn, Fe, and Zn) in solution, increasing their availability to plants (Jauregui and Reisenauer, 1982; Dinkelaker et al., 1989; Rengel and Romheld, 2000). Treeby et al (1989) reported that, in comparison with the complex-forming capability of citrate, amino acids found in the root exudates had limited ability to complex with metal ions. In contrast to amino acids, the simulation involving citric acid predicted that 99% of the citrate in solution was complexed with Al instead of a Ca, Mg, Fe, Mn, Cu, or Zn complex. Compounds that act as phytosiderophores also are known to be released by some plants in response to Zn deficiency and are capable of mobilizing Cu, Mn, and Zn from soils. Plus, they are effective chelators of Ni (Murakami et al., 1989) and Co (Ma and Nomoto, 1993).

Plant roots readily absorb organo-metallic complexes in the soil solution (Jones et al., 1994). Chairidchai and Ritchie (1993) observed that the metal-ligand complex had a residence time in soil of between 17 and 48 hours. Exudates produced in sterile hydroponic and field cultures might be fractionated into low molecular weight soluble and high molecular weight colloidal (such as mucilage) fractions. Both root exudate fractions were able to bind metals and the binding capacity of the soluble fraction was greater than that of the mucilage. The affinities for metal binding were Cu(II) > Pb(II) > Zn(II) and Pb(II) > Cu(II) > Cd(II) > Zn(II) for the soluble and the colloidal fractions, respectively. Mucilages at root surface were thought to limit the penetration of heavy metals into the root, playing the role of a selective barrier. Conversely, soluble exudates produced along the elongation zone of the root initiated the complexing of metals and, thus, enhanced their solubility and absorption by the root. The solubility of heavy metals in root exudates was dependent on the ligands and metals involved (Morel et al., 1987).

The behavior of root exudates that dissolved Zn was identical to the phytosiderophores that were released by plants cultured under Fe deficiency conditions. Marschner et al. (1986) and Takagi et al. (1988) showed that root exudates of wheat plants cultured under Zn- and Fe-deficient conditions mobilized Zn and Fe, respectively, at comparable rates. Similar results were observed for barley plants. The release of root exudates in plants cultured under Zn-deficient conditions followed a distinct diurnal rhythm similar to the release of phytosiderophores by plants cultured under Fe-deficient conditions. They identified 2'-deoxymugineic acid in root exudates of both Zn- and Fe-deficient wheat plants and epi-3-hydroxymugineic acid in root exudates of both Zn- and Fe-deficient barley plants. These were typical phytosiderophores released by wheat and barley under Fe-deficient growing conditions (Romheld and Marschner, 1990). The most important Zn-exudate complexes were Zn-oxalate and Zn–aspartate for pines. In the root exudates of barley, there was little complex reaction between Zn and amino acids (Inskeep and Comfort, 1986).

The role of root exudates in enhancing solubility of soil Mn or Cu has been reported (Nielsen, 1976; Godo and Reisenauer, 1980; Jauregui and Reisenauer, 1982; Merckx et al., 1986; Mench et al., 1988a, 1988b; Uren and Reisenauer, 1988; Linehan et al., 1989). Godo and Reisenauer (1980) found carboxylic acids in the rhizosphere and claimed that malate was responsible for increasing solubility of soil Mn by reducing Mn(IV) to Mn(II) and by complexing the released Mn(II) ions. Jauregui and Reisenauer (1982) suggested that the reduction of Mn(IV) by malate also promoted reactions of other organic acids with Fe. Root exudate-extracted Cd concentrations from the nutrient solutions or soils were in proportion to Cd concentrations in plants. The extent of Cd extracted by root exudates of tobacco (*N. tabacum*) > tobacco (*N. rustica*) > corn was similar to the order of Cd's bioavailability to three plants when grown on the soil. The tobacco species were especially effective in accumulating Cd. An increase in Cd solubility in the rhizosphere of apical root zones due to root exudates was likely to be an important cause of the relatively high Cd accumulation in tobacco plants. Root exudates of corn solubilized mainly Cu or Mn, depending on the Cu/Mn ratio in soil samples (Mench and Martin, 1991). The data suggested that citric acid, glycine, and glucose were not suitable as extractants of metals in simulating the actions of roots. It was observed that Cu-oxalate and Cu-leucine were common in the rhizosphere. For pines, oxalate complexes were prevalent at soil pH <5.4 and leucine complexes were more important at higher pH values. The most important amino acid complexes were Cu-aspartate, Cu-histidine, Cu-glutamine, and Cu-serine for barley (Inskeep and Comfort, 1986).

Krishnamurti et al. (1997) demonstrated that soil Cd might be mobilized by low molecular weight organic acids: acetic, citric, oxalic, fumaric, and succinic acids, which were often found in root exudates. At the same extractant concentration, the release of soil Cd followed the order: fumaric acid > citric acid > oxalic acid > acetic acid > succinic acid. The rate of Cd release might be expressed as:

$$Q_t = D\, t^{1/2} + K \qquad \text{[Eq. 3.7.4-1]}$$

where Q_t is the amount of Cd released in time t, D is the overall dissolution rate coefficient for Cd and K is a constant (Evans and Jurinak, 1976). D and K for metal solid phases in a soil may be determined by kinetic experiments. Krishnamurti et al. (1997) reported that amounts of soil Cd mobilized by organic acids were proportional to the kinetics of Cd released, represented by the D of the soil.

The rate of ion uptake (k) or the rate of ion leakage (k') per unit of fresh weight of root was calculated from changes in the amounts of the ions in the solutions before and after incubation. The index (R) for ion uptake inhibition by allelochemicals was:

$$R = 100\,(k_0 - k) / k_0 \qquad \text{[Eq. 3.7.4-2]}$$

where k_0 is k for the control (Yu and Matsui, 1997).

The fixation or release of heavy metals can be determined by the equilibrium reaction:

$$S_a\,\text{Me(soil)} + H^+\,\text{(sol)} \leftrightarrow S_a\,\text{H (soil)} + \text{Me (sol)} \qquad \text{[Eq. 3.7.4-3]}$$

where S_a represents the different compartments of soil responsible for the fixation of the trace metal Me. This explains the effect of acidity on the mobility of metal (Unger and Allen, 1988). Based on the surface complexation model, Furrer and Stumm (1986) indicated that surface complexes of any ligand should be assumed as chemical species and used directly in dissolution rate laws. The dissolution rate can be expressed:

$$R_H = k_H\,\{>MOH_2^+\}^n = k_H\,(C_H^s)^n \qquad \text{[Eq. 3.7.4-4]}$$

where R_H is the dissolution rate, $\{>MOH_2^+\}^n$ is the concentration of protonated hydroxo groups, and C_H^s is the concentration of adsorbed protons.

CHAPTER 4.0

EFFECTS OF BIOSOLIDS TREATMENT ON PLANT GROWTH

Plant growth is affected by a variety of factors, such as the availability of nutrients and water in the culturing media, presence of potentially toxic substances, conditions of the environment, and physiological attributes of plants. Biosolids application may alter some of these factors in the growing media. These factors also may affect the secretion of exudates and in turn affect plant uptake of biosolids-borne metals.

In this section, we summarize results and discuss the effects of biosolids on plant growth and biomass yields. All plants used in the experiments responded positively to the treatment of biosolids, and the biomass harvests from the nutrient solution irrigated sand media treated with biosolids were significantly higher than those harvested from media with sand alone. The effects of individual factors are summarized in the following sections.

4.1 Stage of Plant Development

The rate of growth for plants (in terms of dry matter accumulation) varied with the physiological stage of plant development. Generally, the rate of growth followed an "S"-shaped curve. Following planting, the plants underwent a short acclimation period, followed by an accelerated growth period. When plants approached maturity, growth slowed rapidly. The biomass of a plant increases in proportion to its rate of growth. This trend was observed in an experiment with corn grown on MWRDGC 1975 biosolids-treated sand media for 16 weeks. The experimental procedures are described in Section 2.1.

During the first two weeks of the growth period, the corn plants receiving nutrient solution and biosolids-treatment grew at comparable rates (Figure 4-1). Afterwards, the rates of growth for plants on the biosolids-treated media accelerated exponentially for the next 10 weeks. For the same period of time, the corn plants grown on the media irrigated with standard nutrient solution grew at considerably slower rates and accumulated significantly lesser amounts of biomass. In both treatments, the biomass did not change significantly during the final stage of growth (12–16 weeks). At the end of the experiment, the corn plants receiving the biosolids-treatments were 4–5 times greater in biomass than those that received the standard nutrient solution treatments. The root biomass was approximately 20–30% of the shoot biomass and the trends of root biomass formation with respect to time were similar to those of shoots (Figure 4-2). The difference in plant growth could not be explained entirely by the deficiency in nutrients, as the nutrients in the standard nutrient solutions were routinely replenished throughout the growing season.

In plant growth experiments involving biosolids, investigators often observe enhanced plant growth. Plant growth hormones, such as the auxins, have been found in biosolids (Rudolfs and Ghem, 1942). For the present research, analyses of the biosolids-treated media detected the precursor of the microbial-induced plant growth regulator, tryptophan. It is possible, therefore, that the enhanced growth in the biosolids-treated media might be related to the presence of plant growth regulators produced by microbial reactions. This phenomenon appeared to be plant species-specific and cannot be confirmed with results of this experiment. More likely, the biosolids provided additional nutrients, especially N, that significantly increased the nutrient intensity in the rhizosphere and thus accelerated the plant uptake of nutrients. This issue will be revisited in later discussions of plant species effects (Section 4.2) and biosolids sources effects (Section 4.4).

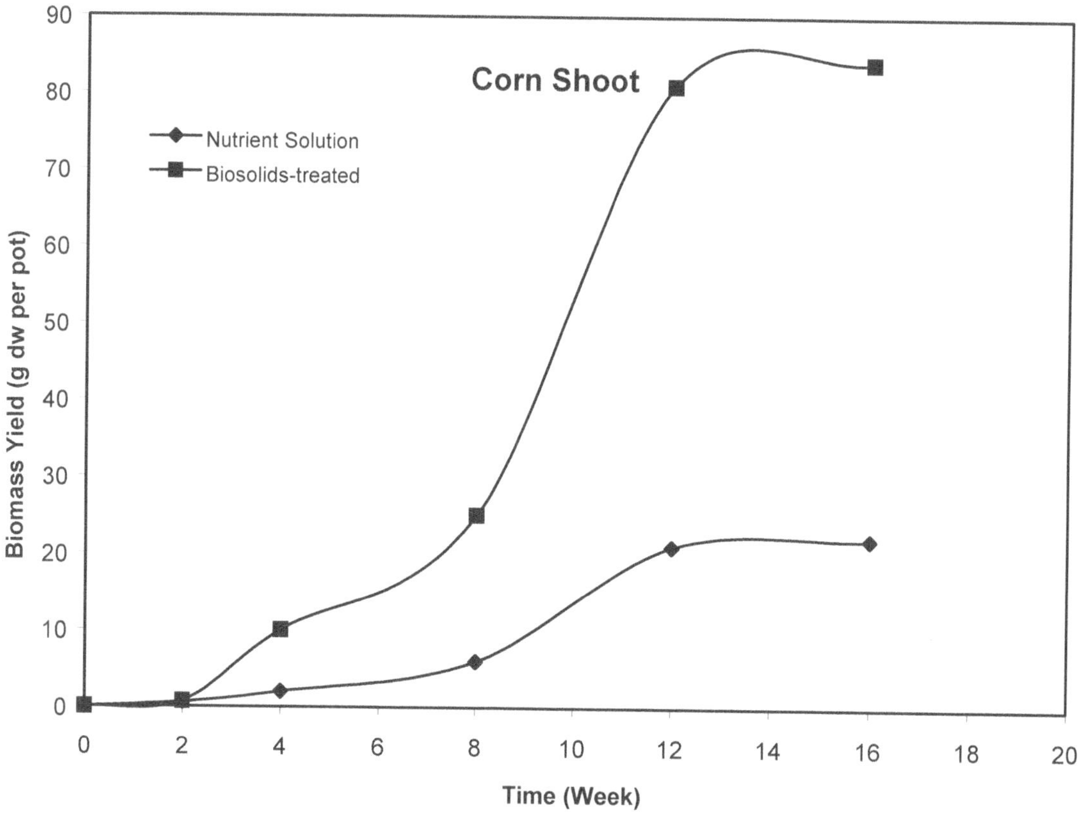

Figure 4-1. Shoot Biomass Yields of Corn Grown on Sand Media Treated with MWRDGC 1975 Biosolids and Nutrient Solution.

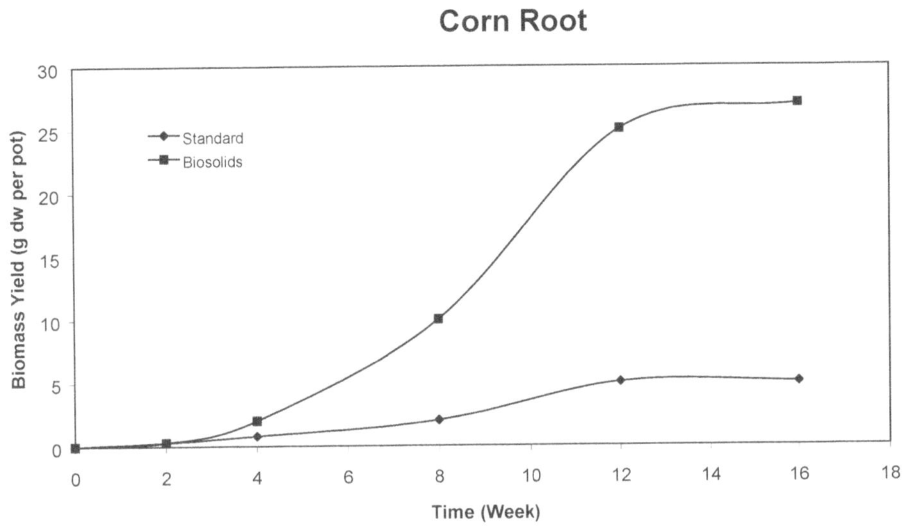

Figure 4-2. Root Biomass Yields of Corn Grown on Sand Media Treated with MWRDGC 1975 Biosolids and Nutrient Solution.

4.2 Plant Species

Three monocotyledonous species (Durum wheat, Sudan grass, and corn) and three dicotyledonous species (canola, chickpea and Swiss chard) were grown for four weeks on sand media treated with biosolids plus nutrient solution; phosphorus rock plus nutrient solution; and nutrient solution only. The MWRDGC 1975 biosolids was used for the biosolids treatments. Roots of some plant species responded to P-deficiency by excreting organic compounds that convert P from forms unavailable to plants to forms available to plants. This treatment was included to identify species that responded to phosphorus rock, a less readily available P source for plant growth.

Response to the experimental treatment differed among plant species (Figure 4-3). The monocotyledonous species (monocot, hereafter) responded positively to the biosolids treatment and the biomass yields were significantly higher that their respective controls as represented by the standard sand media treatment. However, the monocot species failed to utilize effectively the P provided by phosphorus rock. The biomass yields of wheat, Sudan grass, and corn grown on phosphorus rock-treated media were no more than 20% of that in the control treatment.

For the dicotyledonous (dicot, hereafter) species (chickpea, canola and Swiss chard), the biomass yields of the biosolids-treatment, phosphorus rock-treatment, and the control were comparable. The additional nutrients in the growing media did not further enhance the plant growth. Dicot species effectively utilized the P provided by the phosphorus rock. Their biomass yields in standard, biosolids, and phosphorus treatments were comparable and not significantly different.

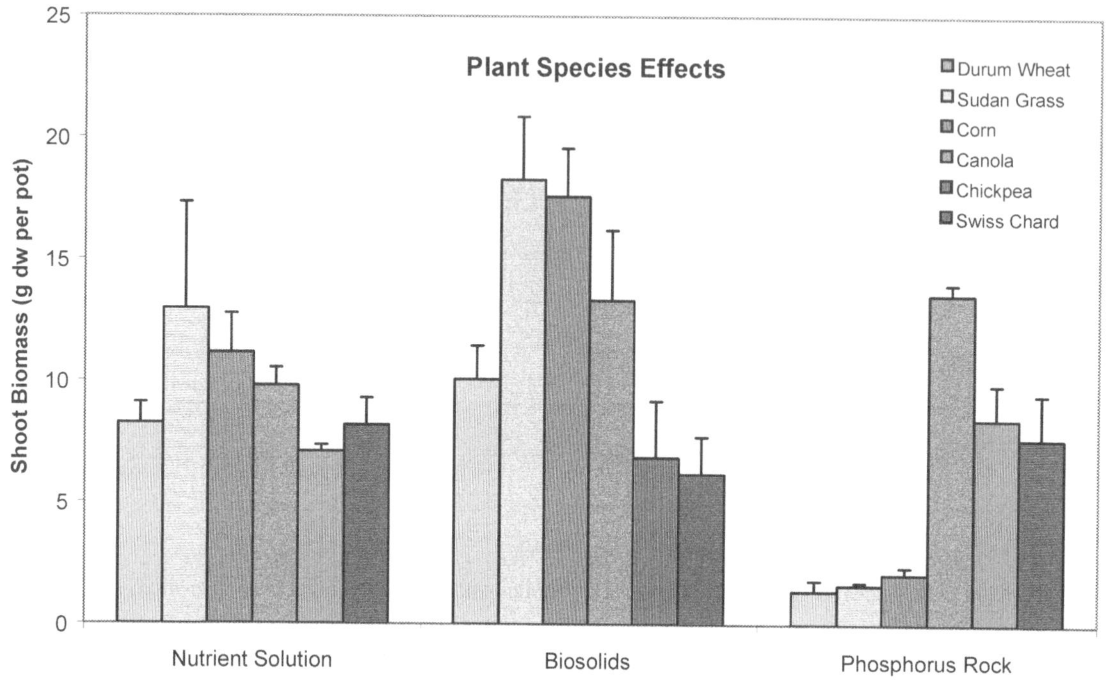

Figure 4-3. Shoot Biomass of Six Species Grown for Four Weeks on Sand Media Treated with MWRDGC 1975 Biosolids, Phosphorus Rock, and Nutrient Solution (Average ± SD).

4.3 Variety

Three varieties of wheat (*Tricutum aestivum L.* cv. Durum, *Tricutum turgidum* cv. Chinese spring, and cv. Yecora Rojo) were grown on biosolids-treated media for 12 weeks and the biomass accumulation was measured at Weeks 4, 8, and 12. The rates of growth of plants in standard nutrient solution and the biosolids treatments were similar for the first eight weeks. Afterwards, the growth of all three varieties accelerated exponentially and the rates of growth were different for the three cultivars (Figure 4-4). The biomasses harvested at the12th week of growth were significantly different ($p < 0.01$) and the biomass of Chinese spring > Durum > Yecora Rojo. The differences in the growth pattern for corn grown on biosolids-treated media and the standard nutrient solution media persisted in these three varieties of wheat. At the 12th week, the biomasses of plants grown on the biosolids-treated media were, on the average, twice that of comparable varieties grown on the standard nutrient solution media. The accumulation of root biomass followed a similar pattern. The higher biomass produced by Chinese spring resulted from its late maturity, as the other two varieties of wheat flowered about two weeks earlier. The varietal differences in plant development and biomass yield were inherent to their genetic dissimilarity.

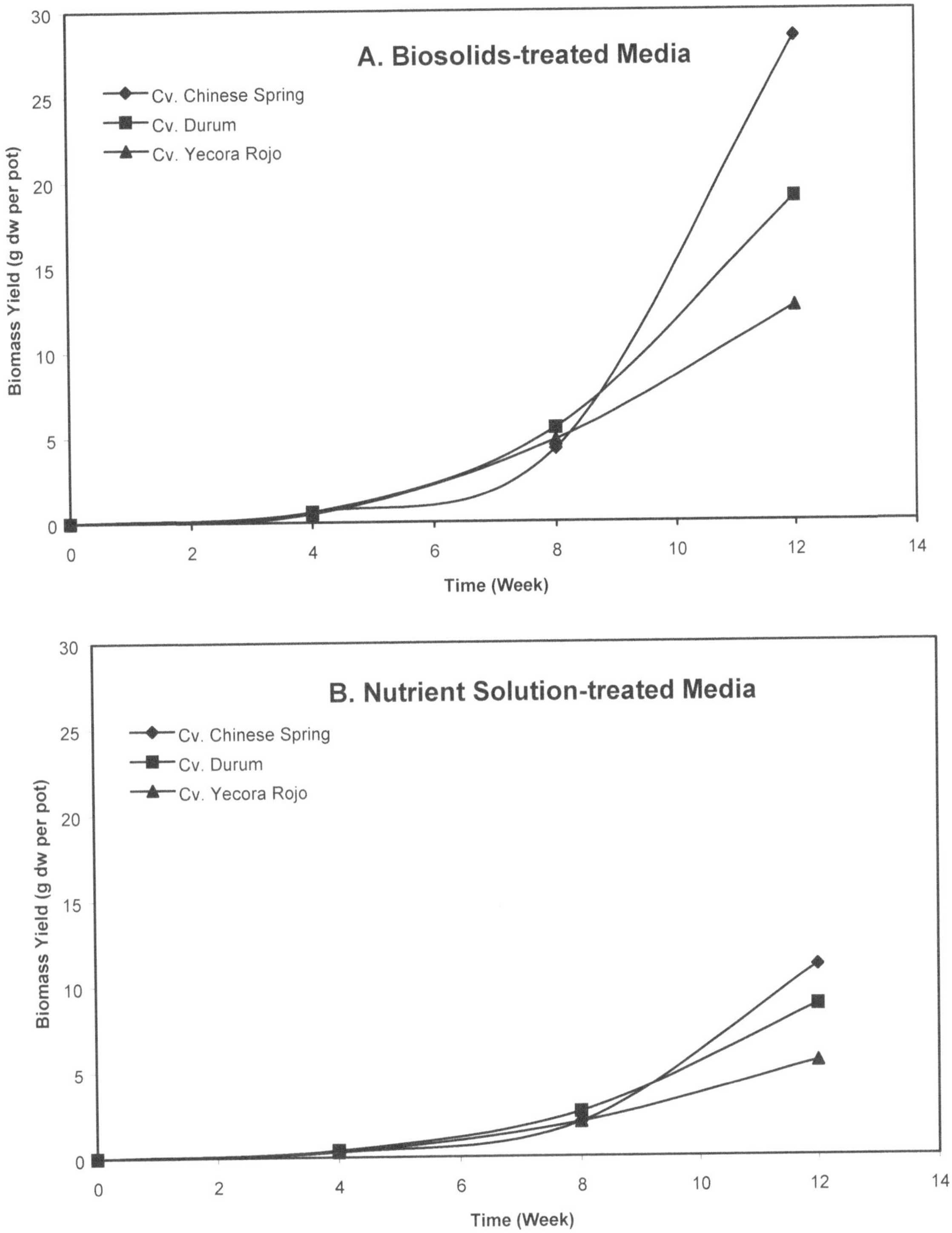

Figure 4-4. Biomass Yields of Three Wheat Varieties (*Tricutum aestivum L.* cv. Durum, *Tricutum turgidum* cv. Chinese spring, and cv. Yecora Rojo) Grown on Biosolids-Treated Media (A) and Nutrient Solution-Treated Media (B).

4.4 Source of Biosolids

Corn was grown for 12 weeks on sand media treated with biosolids from five sources: MWRDGC 1975; Milorganite, MWRDGC 1999; Los Angeles; and Riverside. The N and metal contents of these five biosolids were different (Table 2-2). Under the same treatment conditions, plants are expected to respond to the biosolids treatment in different manners.

The biomass of corn grown on media treated with biosolids from these five sources followed the same pattern (Figure 4-5). During the 12-week growing period, the biomass of corn increased exponentially and the rates of increase were dependent on the source of biosolids but were considerably faster than for corn irrigated with the nutrient solution alone. At the 12th week, corn grown on biosolids-treated media resulted in 5-11 times greater above-ground biomass than the control (grown on standard nutrient solution treated sand media). Statistical testing indicated that the biomass of corn grown on Riverside = Los Angeles = MWRDGC 1999 > Milorganite > MWRDGC 1975 > standard nutrient solution only, at $p < 0.05$. The total biomass accumulation for biosolids treatments was in proportion to the N contents of the biosolids, except Milorganite, whose N availability likely was affected by the heat treatment. It appeared that the biosolids treatments created reservoirs of additional nutrients in the rhizosphere. Nutrients in this reservoir were slowly released over time to create a constant N pool in the rhizosphere and further accelerated the plant growth while the nutrient levels in the media irrigated with nutrient solution alone might have fluctuated considerably between the irrigations.

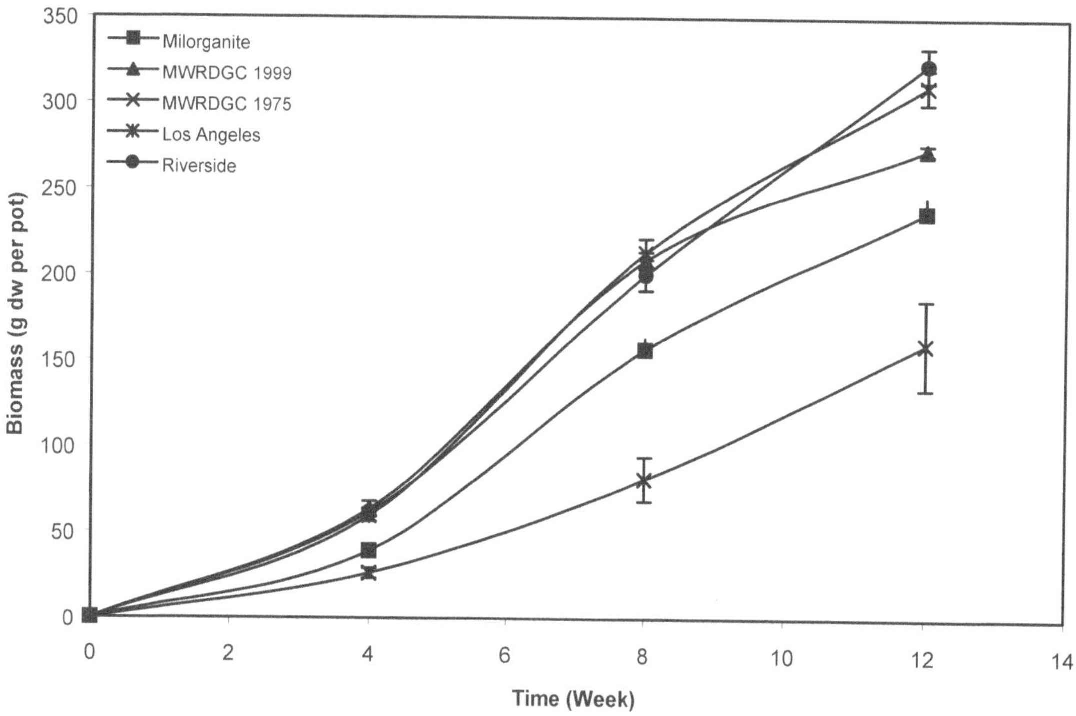

Figure 4-5. Shoot Biomass of Corn Grown on Sand Media Treated with One of Five Sources of Biosolids (Average ± SD).

4.5 Conclusions

Plants generally responded positively to biosolids incorporated into the growth media. The rates of growth and the increases in biomass yields, however, were plant species-dependent and within a species, variety-dependent. The sources of biosolids also affected the growth rate, as the nutrient content and availability in biosolids from each source were not the same. In general, the biomass yields of plants grown on biosolids-treated media were proportional to the total nitrogen content of the biosolids. The metal concentrations of the biosolids or the metal loading in the rhizosphere did not affect the performance of plant grown on biosolids-treated soils.

CHAPTER 5.0

Plant Uptake of Biosolids-Borne Metals

5.1 Metal Uptake by Corn Grown on MWRDGC 1975 Biosolids-Treated Media

In this experiment, corn was grown for 16 weeks on sand media irrigated with nutrient solution and on nutrient solution irrigated sand media that were amended with 20 Mg ha^{-1} of MWRDGC 1975 biosolids. When corn plants were grown on the MWRDGC 1975 biosolids-treated sand media, the metal concentrations in the plant roots were significantly higher than or, in few cases, not significantly different from those present in respective plant shoots. Plants grown in biosolids-treated media had significantly higher concentrations of metals than those of the standard nutrient solution only sand media at all stages of growth (Figures 5.1–5.6). For metals that were either not readily available to the plant (such as Pb and Cr) or present in limited amounts (such as Cd, which was not supplied in standard medium), the amount of metals absorbed appeared to be limited by their supply in the growing media. After planting, the supply was quickly exhausted and their concentrations in plant tissue decreased as biomass accumulated over time due to mass dilution (Figures 5-1 and 5-5). This is an indication that, under the experimental conditions, metal availability in growing media did not keep pace with the rate of biomass accumulation. These observations may be illustrated by a comparison of the average Cr, Cu, Ni and Pb concentrations of plants during 0-4 and 4-16 weeks of the growing period (Table 5-1). In all cases, the average Cr and Pb concentrations of plants during the first four weeks were significantly higher than the average concentrations of those during 4-16 weeks. As the available metals in the growing media diminish, the mass dilution resulted in a gradual reduction of plant metals concentrations from the 0-4 and 4-16 weeks.

Table 5-1. Average Cr, Cu, Ni, and Pb Concentrations of Corn Plants Harvested from Sand Media Treated with MWRDGC 1975 Biosolids and Standard Nutrient Solution at 0-4 and 4-16 Weeks.

Element	Plant Part	Treatment	Average Concentration (μg g^{-1})[a]	
			0–4 Weeks	4–16 Weeks
Cr	Shoot	Nutrient Solution	1.20 (0.11) a	0.83 (0.03) b
		MWRDGC 1975	2.70 (0.49) a	1.43 (0.06) b
	Root	Nutrient Solution	1.80 (0.84) a	1.17 (0.07) b
		MWRDGC 1975	13.5 (0.84) a	6.00 (1.42) b
Pb	Shoot	Nutrient Solution	5.50 (0.65) a	3.33 (0.27) b
		MWRDGC 1975	13.5 (1.01) a	9.33 (0.71) b
	Root	Nutrient Solution	5.50 (0.65) a	2.33 (0.37) b
		MWRDGC 1975	27.1 (1.95) a	11.1 (0.27) b
Cu	Shoot	Nutrient Solution	12.0 (0.55) a	6.67 (0.13) b
		MWRDGC 1975	17.0 (0.55) a	12.7 (0.80) b
	Root	Nutrient Solution	22.0 (1.45) a	18.3 91.20) b
		MWRDGC 1975	63.5 (6.70) a	40.3 (3097) b
Ni	Shoot	Nutrient Solution	14.5 (0.90) a	9.00 (0.57) b
		MWRDGC 1975	44.0 (4.65) a	23.3 (2.01) b
	Root	Nutrient Solution	19.5 (2.15) a	12.7 (0.60) b
		MWRDGC 1975	276 (49.5) a	149 (41.7) b

[a]Values in the parentheses denote the standard deviation. Letters following the value indicate statistical significance, where values followed by the same letter were not statistically significant at $p<0.05$.

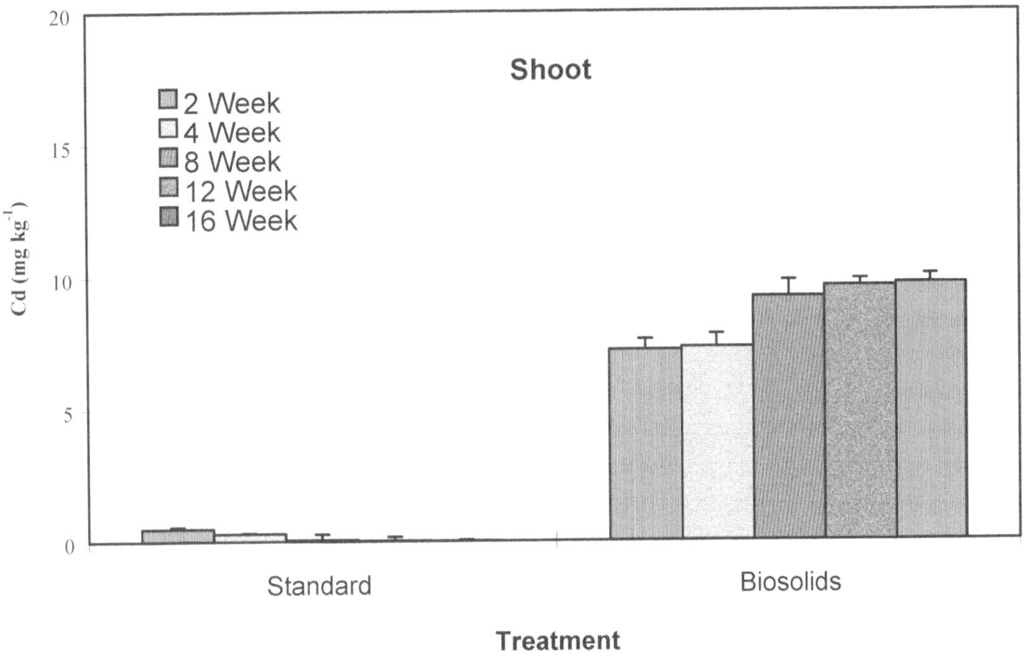

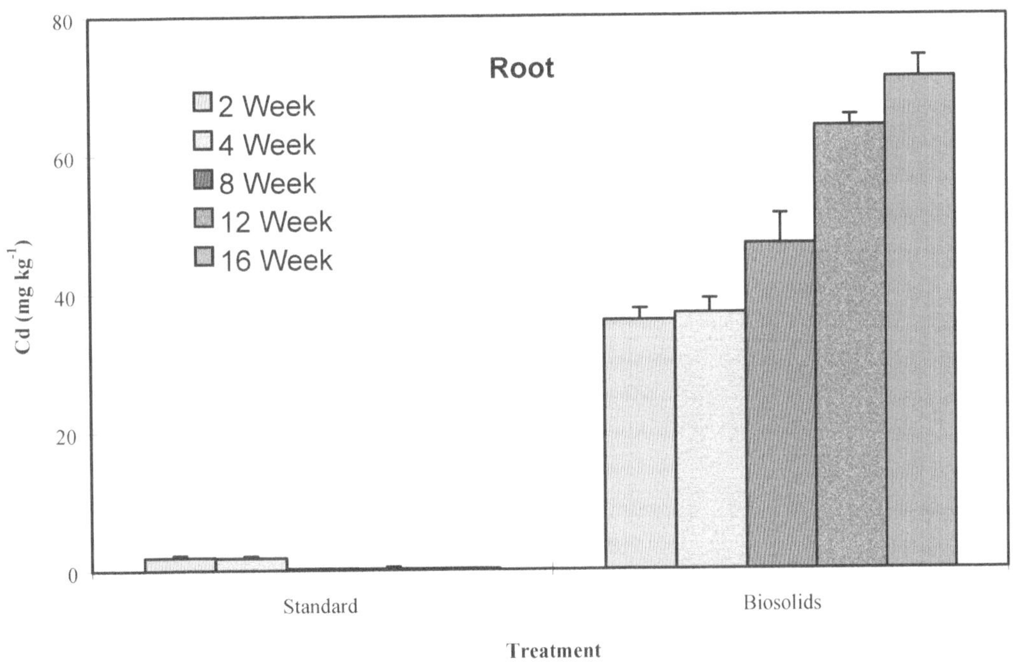

Figure 5-1. Cd Concentration of Shoot and Root of Corn Grown on Sand Media Treated with Standard Plant Nutrient and MWRDGC 1975 Biosolids.

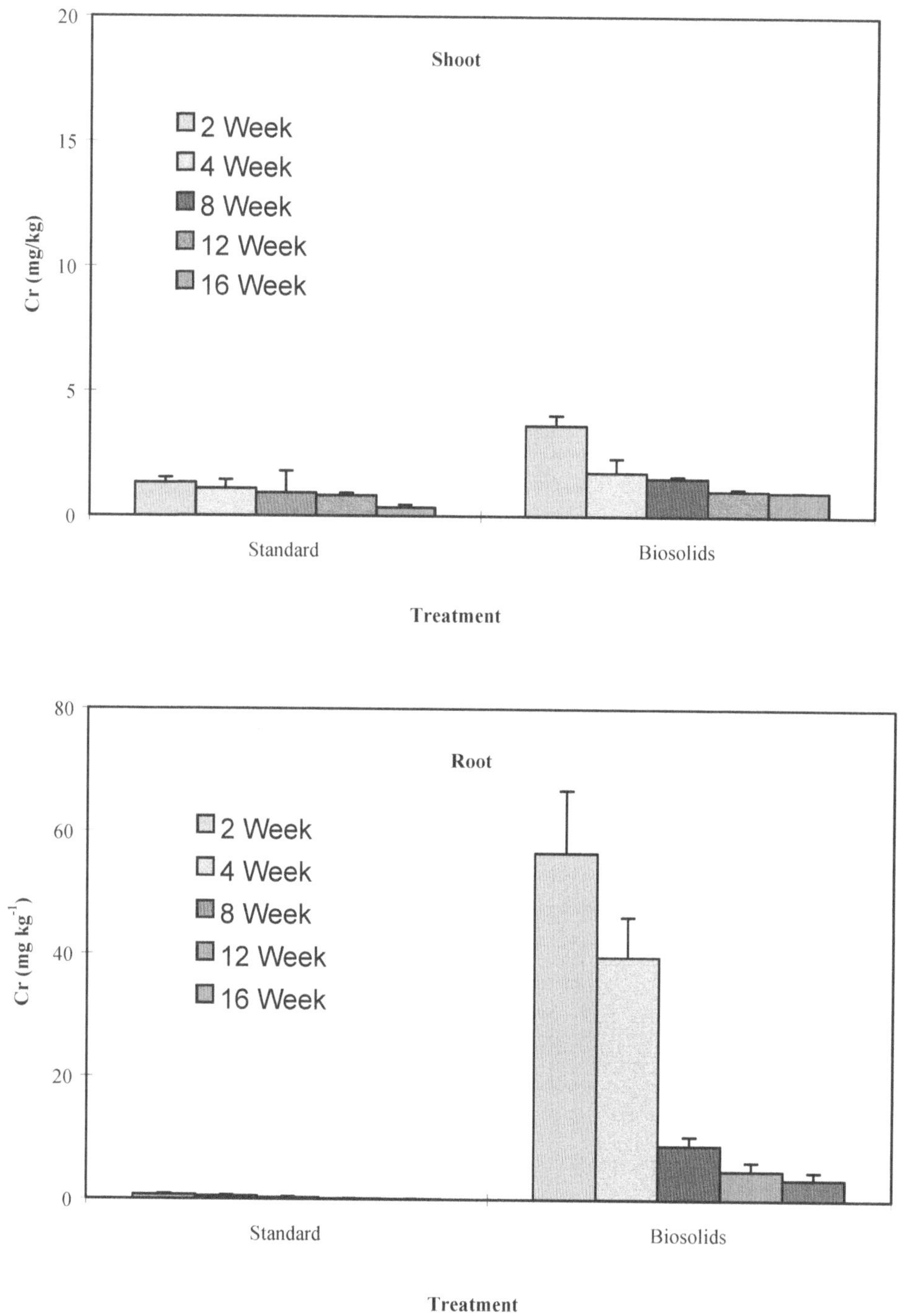

Figure 5-2. Cr Concentration of Shoot and Root of Corn Grown on Sand Media Treated with Standard Plant Nutrient and MWRDGC 1975 Biosolids.

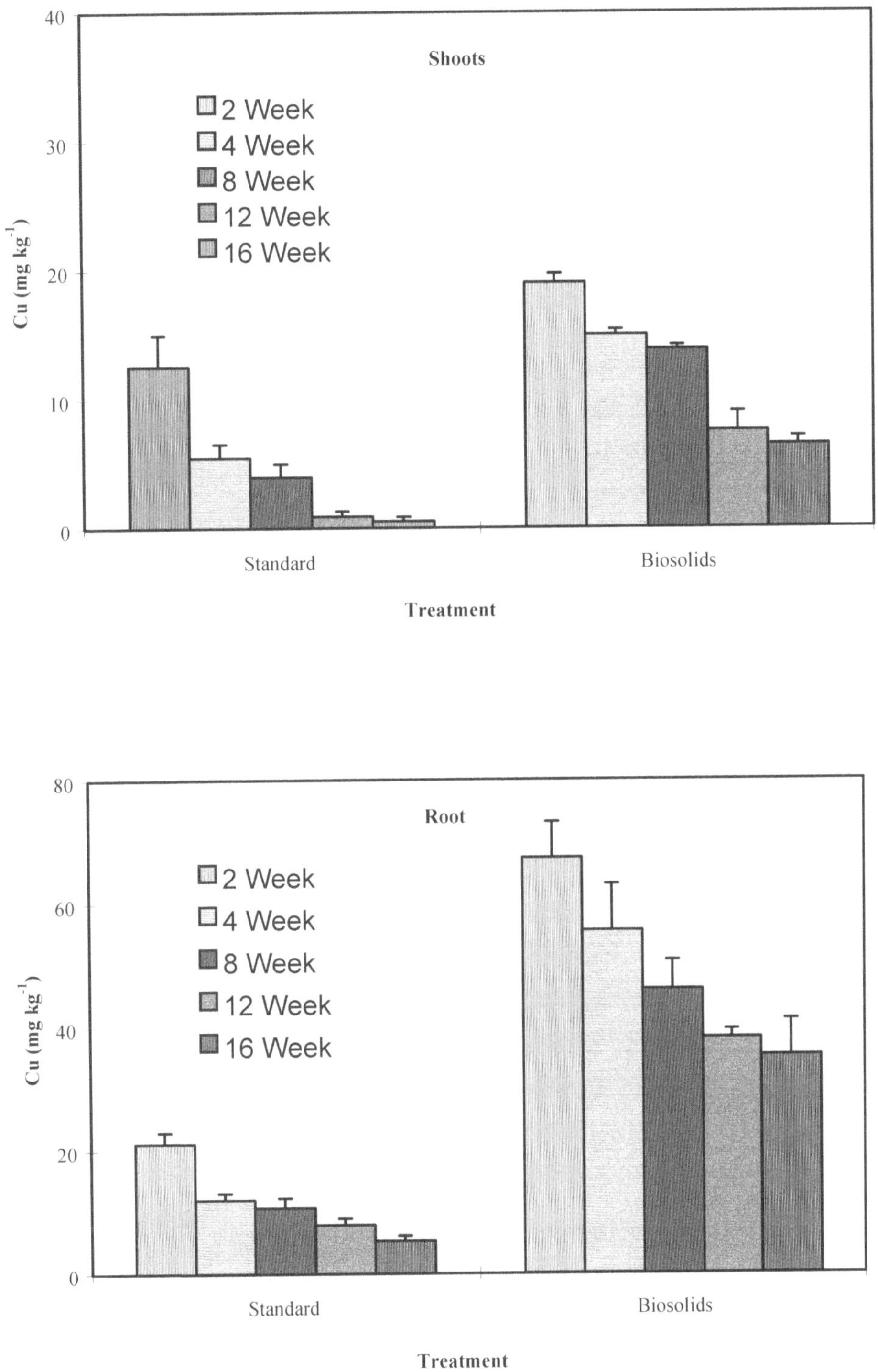

Figure 5-3. Cu Concentration of Shoot and Root of Corn Grown on Sand Media Treated with Standard Plant Nutrient and MWRDGC 1975 Biosolids.

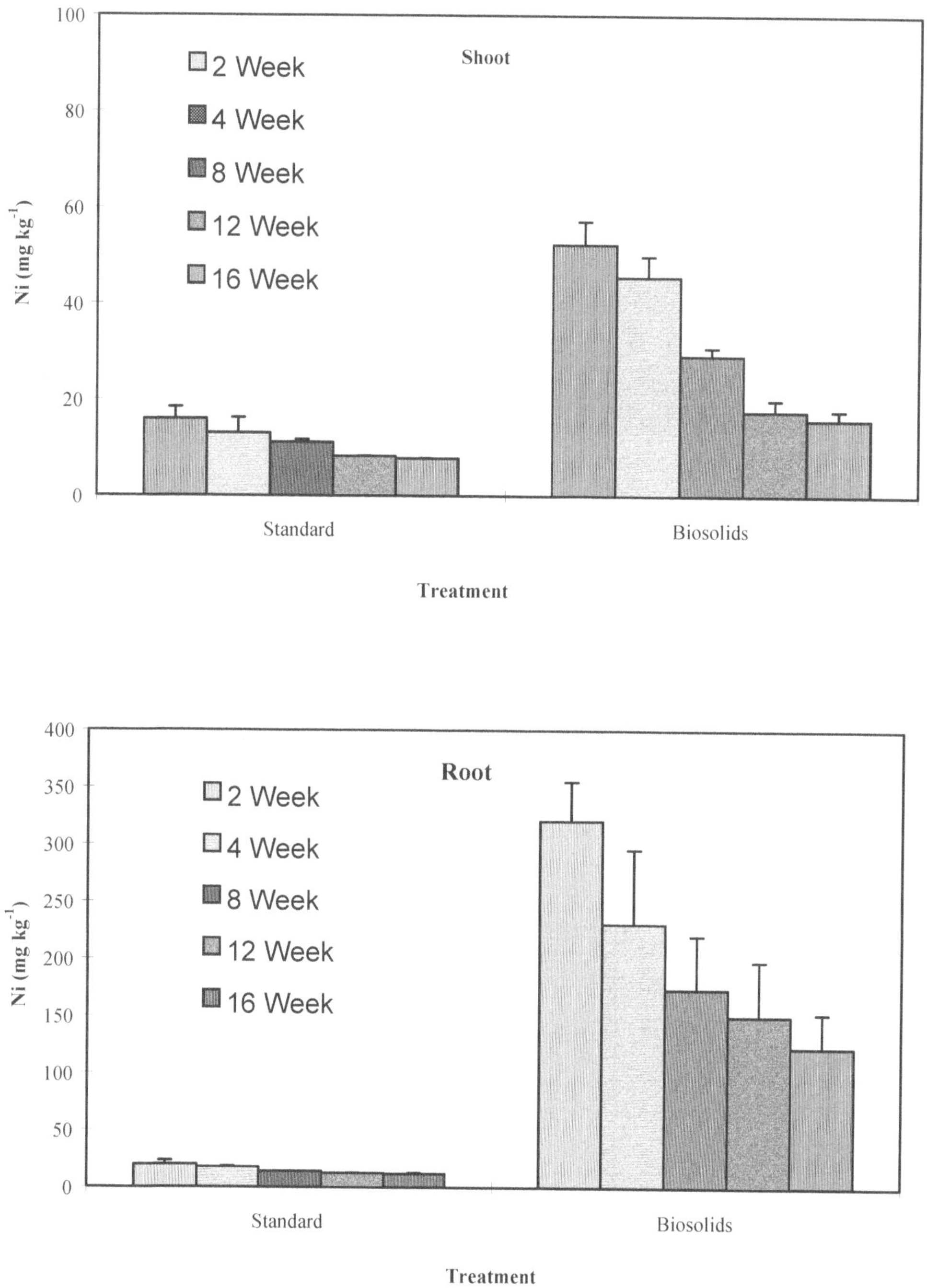

Figure 5-4. Ni Concentration of Shoot and Root of Corn Grown on Sand Media Treated with Standard Plant Nutrient and MWRDGC 1975 Biosolids.

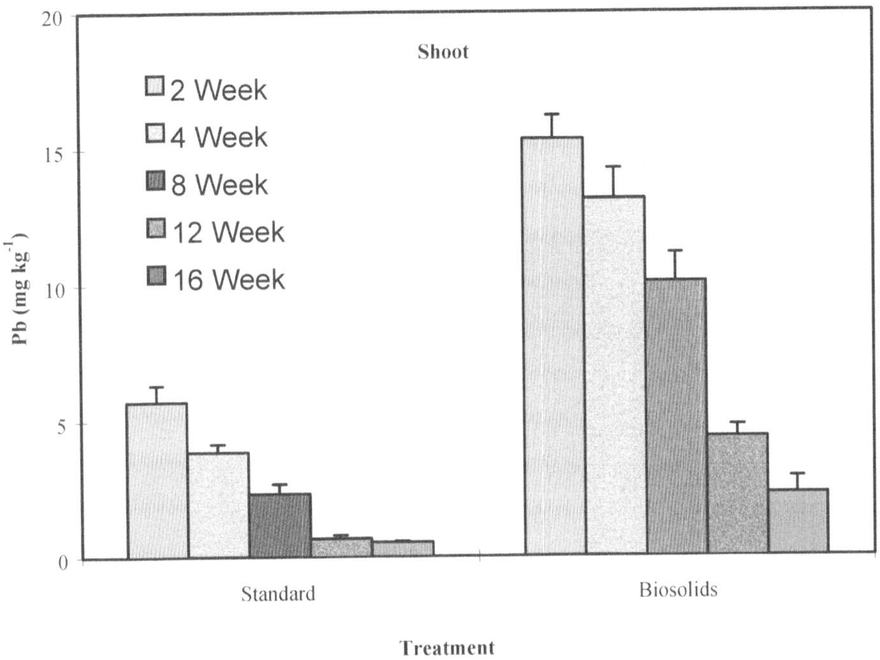

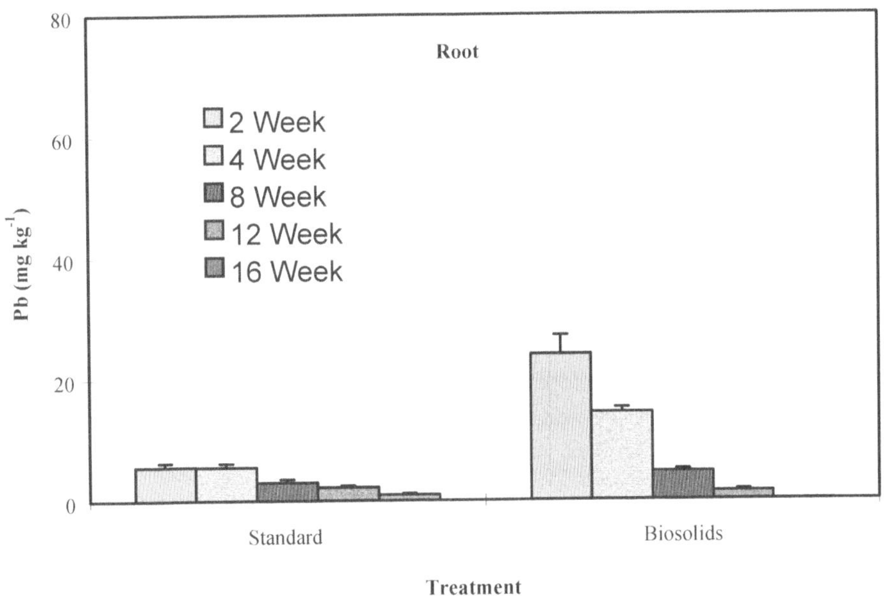

Figure 5-5. Pb Concentration of Shoot and Root of Corn Grown on Sand Media Treated with Standard Plant Nutrient and MWRDGC 1975 Biosolids.

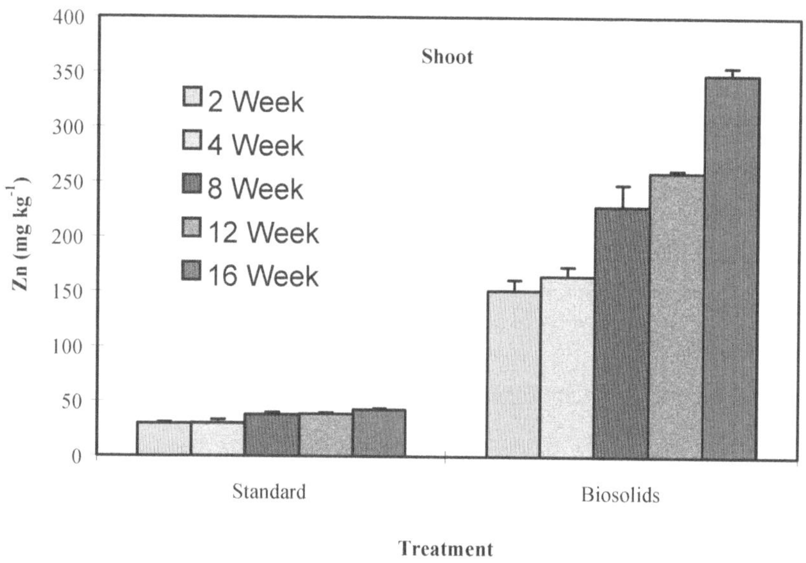

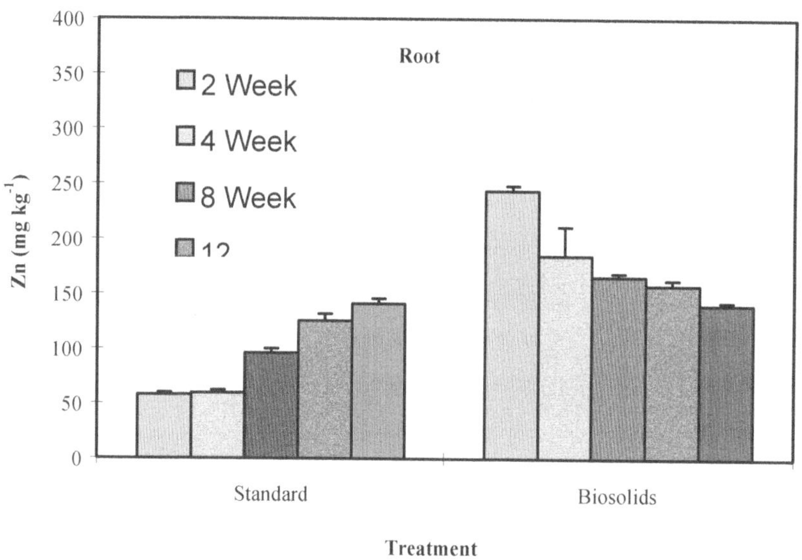

Figure 5-6. Zn Concentration of Shoot and Root of Corn Grown on Sand Media Treated with Standard Plant Nutrient and MWRDGC 1975 Biosolids.

For metals that were readily absorbed by plants, the amounts absorbed and the plant tissue concentrations were dependent on the amounts present in the growing media. Significant amounts of Cu and Ni were absorbed by plants grown in both the standard and the biosolids-treated media (Figures 5-3 and 5-4). In the standard media, amounts of Cu and Ni available were determined by the micro-nutrients added into the nutrient solution and their supply slowly diminished as the plants absorbed them. As a result, the plant tissue concentrations for the 0-4 and 4-12 weeks followed the same pattern as those of Cr and Pb (Table 5-1). Cd and Zn were readily available to plants and were in abundant supply in the growing media (such as Cd in biosolids-treated media or Zn in both standard and biosolids-treated media), their concentration in the plant tissue either continued to rise in the shoots grown on both standard and biosolids-treated media over time or remain the same with plant development (Figures 5-1 and 5-6). For example, Zn concentration in the shoot and root of corn grown on the standard media and in the shoot of the biosolids-treated media increased with time following planting. Zn concentrations in the roots of corn grown on biosolids treated media, however, either remained the same or decreased slightly with time following planting (Figure 5-6). In plant shoots, accumulations of Cd, Ni, and Zn were especially notable, because their concentrations in the plant tissue that received the biosolids treatment were several times higher than those that received the standard nutrient solution treatment (Figures 5-1, 5-4, and 5-6). The Cd and Zn concentration of the 0-4 and 4-16 weeks plant tissues are summarized in Table 5-2.

Table 5-2. Average Cd and Zn Concentrations of Corn Plants Harvested from Sand Media Treated with MWRDGC 1975 Biosolids and Standard Nutrient Solution at 0-4 and 4-16 Weeks.

Element	Plant Part	Treatment	Average Concentration ($\mu g\ g^{-1}$)[a]	
			0-4 Weeks	4-16 Weeks
Cd	Shoot	Nutrient Solution	0.38 (0.04) a	0.03 (0.004) b
		MWRDGC 1975	9.50 (1.09) a	7.25 (1.27) b
	Root	Nutrient Solution	1.80 (0.84) a	1.17 (0.07) b
		MWRDGC 1975	13.5 (0.84) b	6.00 (1.42) a
Zn	Shoot	Nutrient Solution	5.50 (0.65) b	3.33 (0.27) a
		MWRDGC 1975	13.5 (1.01) b	9.33 (0.71) a
	Root	Nutrient Solution	5.50 (0.65) b	2.33 (0.37) a
		MWRDGC 1975	27.1 (1.95) a	11.1 (0.27) b

[a]Values in the parentheses denote the standard deviation. Letters following the value indicate statistical significance, where values followed by the same letter were not statistically significant at $p<0.05$.

Based on the mass of metals that was present in the plant growth media and absorbed by plants, the percentages of the biosolids-borne metals in the rhizosphere that were absorbed by plants were calculated (Table 5-3). These percentages are indicators of the phytoavailability of the various metals and, in descending order, their availabilities are Cd > Ni = Zn > Cu = Pb > Cr.

Table 5-3. Metal Uptake by Corn Grown on MWRDGC 1975 Biosolids-Treated Sand Media.

Element	Mass Input (μg per pot)	Mass Output (μg per pot)	Plant Uptake (% input)
Cd	19,198	2,722	14.0
Cr	237,818	263	0.11
Cu	121,829	1,980	1.6
Ni	55,608	5,434	9.8
Pb	77,167	1,052	1.4
Zn	386,076	33,196	8.6

The rate of metal uptake was measured as the amount of metal absorbed per unit biomass increase during the plant-growing period (μg g^{-1} $2wk^{-1}$). For Cd in the standard media where the supply was limited, there was little additional Cd uptake after the first two weeks of growth as the available Cd was exhausted (Table 5-4). When the Cd supply was not limited (i.e., biosolids treatments), the metal uptake rate remained reasonably constant during the active plant developing period of the growing season (0–8 weeks) (Table 5-4). Toward the later part of the growing season, however, there was a tendency for the uptake rates of the roots—but not the shoot—to become elevated as the production of root biomass slowed down while plant absorption of metals continued.

Table 5-4. Rate of Metals Uptake by Corn Plants Grown for Various Time Periods on Nutrient Solution- and MWRDGC 1975 Biosolids-Treated Media.

Element	Time (Weeks)	Shoot Uptake (μg g^{-1} $2wk^{-1}$)		Root Uptake ((μg g^{-1} $2wk^{-1}$)	
		Nutrient Solution	MWRDGC 1975	Nutrient Solution	MWRDGC 1975
Cd	0–2	2.4	7	3.5	37
	2–4	n.d.[a]	7	n.d.	38
	4–8	n.d.	6	n.d.	27
	8–12	n.d.	6	n.d.	87
	12–16	n.d.	7	n.d.	73
Zn	0–2	30	157	58	235
	2–4	30	165	61	168
	4–8	30	158	98	77
	8–12	20	387	127	61
	12–16	63	687	161	31
Pb	0–2	6	15	6	37
	2–4	4	12	3	9
	4–8	1.4	4	n.d.[a]	5
	8–12	0.9	4	n.d.	2
	12–16	n.d.	4	n.d.	5

[a]n.d., not detected

The rate of Zn uptake was considerably higher than that of Cd (Table 5-4). In the standard plant nutrient irrigated sand media, Zn was supplied and the Zn accumulation by corn shoot was steady until the 12th week. Considerably larger amounts of Zn were present in the shoots and roots in the biosolids-treated media. The biosolids-borne Zn was in solid forms and had to dissolve before Zn was available for plant absorption. Yet, the comparable rates of Zn uptake by corn shoot grown on biosolids-treated media were considerably larger than by plants grown on standard sand media. Again, the rate of uptake for shoots remained reasonably constant for the active plant development period, 0–8 weeks. The biomass growth of the plants slowed and then stopped as the plants approached maturity. Apparently, the plant absorption of Zn continued after biomass growth had stopped, indicating that Zn was soluble in water and was transferred into the plant tissue through the mass flow in the water transpiration process. A part of the Zn absorbed at this stage also might have come from translocation from roots to the shoots as the Zn concentration of the root decreased in proportion to Zn concentration increases in the shoots.

In the standard plant nutrient media, Pb was not supplied. Instead, Pb was present primarily as a contaminant in reagents used and as air-borne particulates in the ambient environment. In this circumstance, the available Pb in the growth media would be limited. The rate of Pb uptake appeared to follow a first order reaction model in that the rate of uptake gradually decreased over time as available Pb diminished in the rhizosphere (Table 5-4). In biosolids-treated media, the rate of uptake was faster, but the same uptake pattern was observed. It appeared that the availability of biosolids-borne Pb was limited by very slow dissolution (or diffusion) kinetics. Once the Pb available in the immediate vicinity was exhausted, the rate of Pb absorption decreased considerably. Subsequently, the Pb concentration in the plant tissue decreased due to mass dilution.

5.2 Metal Uptake of Three Wheat Cultivars Grown on Biosolids-Treated Media

In this experiment, three varieties of wheat were grown on sand media irrigated with only the nutrient solution and on sand media amended with 20 Mg ha^{-1} of MWRDGC 1975 biosolids for 12 weeks. Metal concentration in the plant tissues of three wheat varieties (that is, *Tritium aesticum* L. cv Chinese spring, Yecora Rojo, and *Triticum aestivum* L. cv Durum) varied with the treatments, the metal element, the plant part, and the duration of plant growth (Figures 5-7 to 5-12). For Cd, Cr, Cu, Ni, Pb, and Zn, the concentrations in the plant tissue from plants harvested from the MWRDGC 1975 biosolids-treated media were higher than the corresponding varieties grown on standard nutrient-only media. In all cases, metal concentrations of plant tissue were highest at the beginning and decreased over the growing season. The metal concentrations in the plant tissue may be divided according to the rate of plant growth into the early stage (0–4 weeks) and rapid growing stage (4–12 weeks) (Tables 5-5 and 5-6). For all of the metals considered, the differences in average plant tissue concentrations were significant between 0–4 weeks and 4–12 weeks at $p < 0.05$. The range of differences in average plant tissue metal concentrations between the biosolids and the standard plant nutrient-only treatments was much greater than the range of differences between the three wheat cultivars. In fact, the differences in plant tissue metal concentrations were not significantly different between cultivars, at $p < 0.05$.

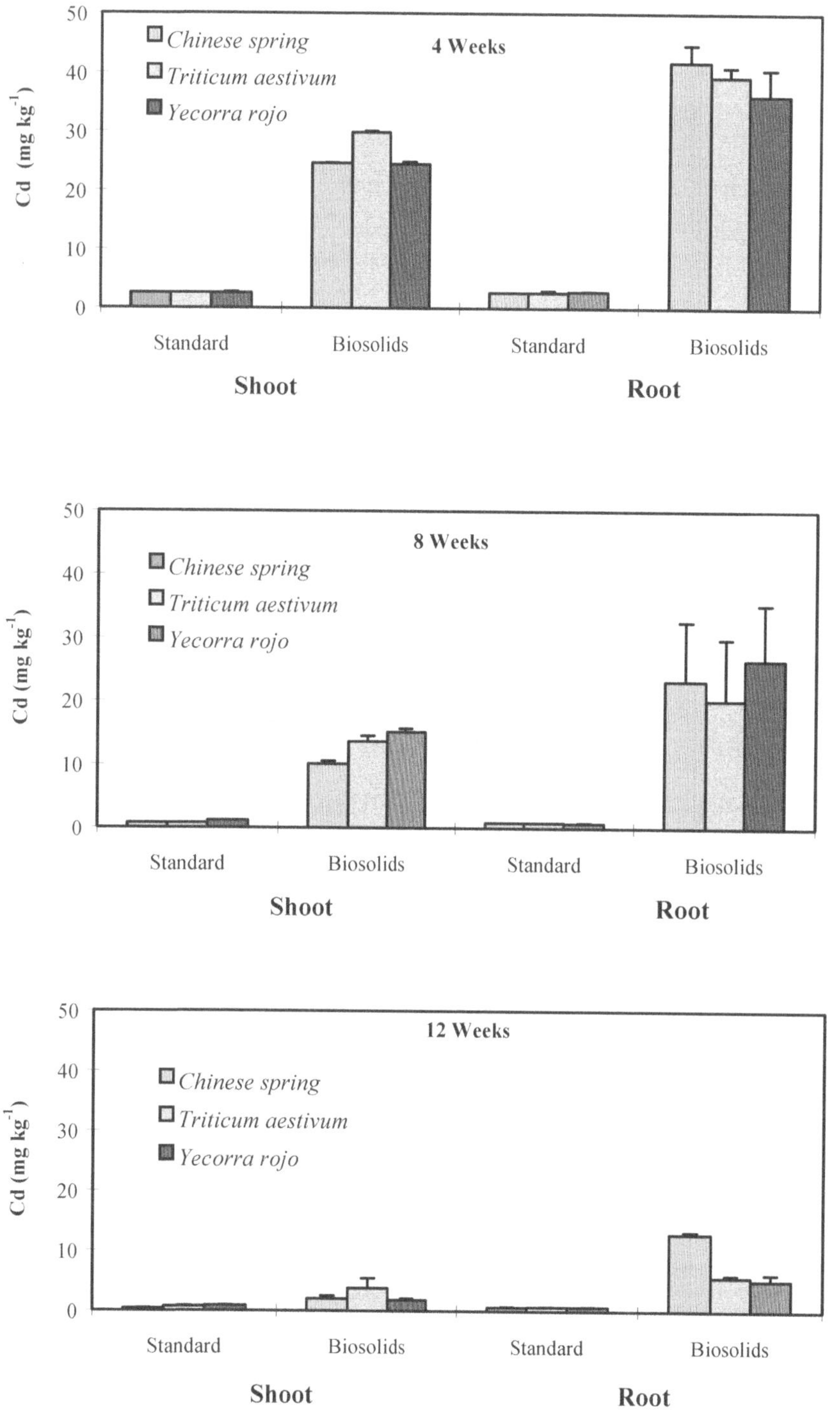

Figure 5-7. Cd Concentrations of Three Wheat Varieties Grown on Standard Nutrient Solution- and MWRDGC 1975 Biosolids-Treated Media.

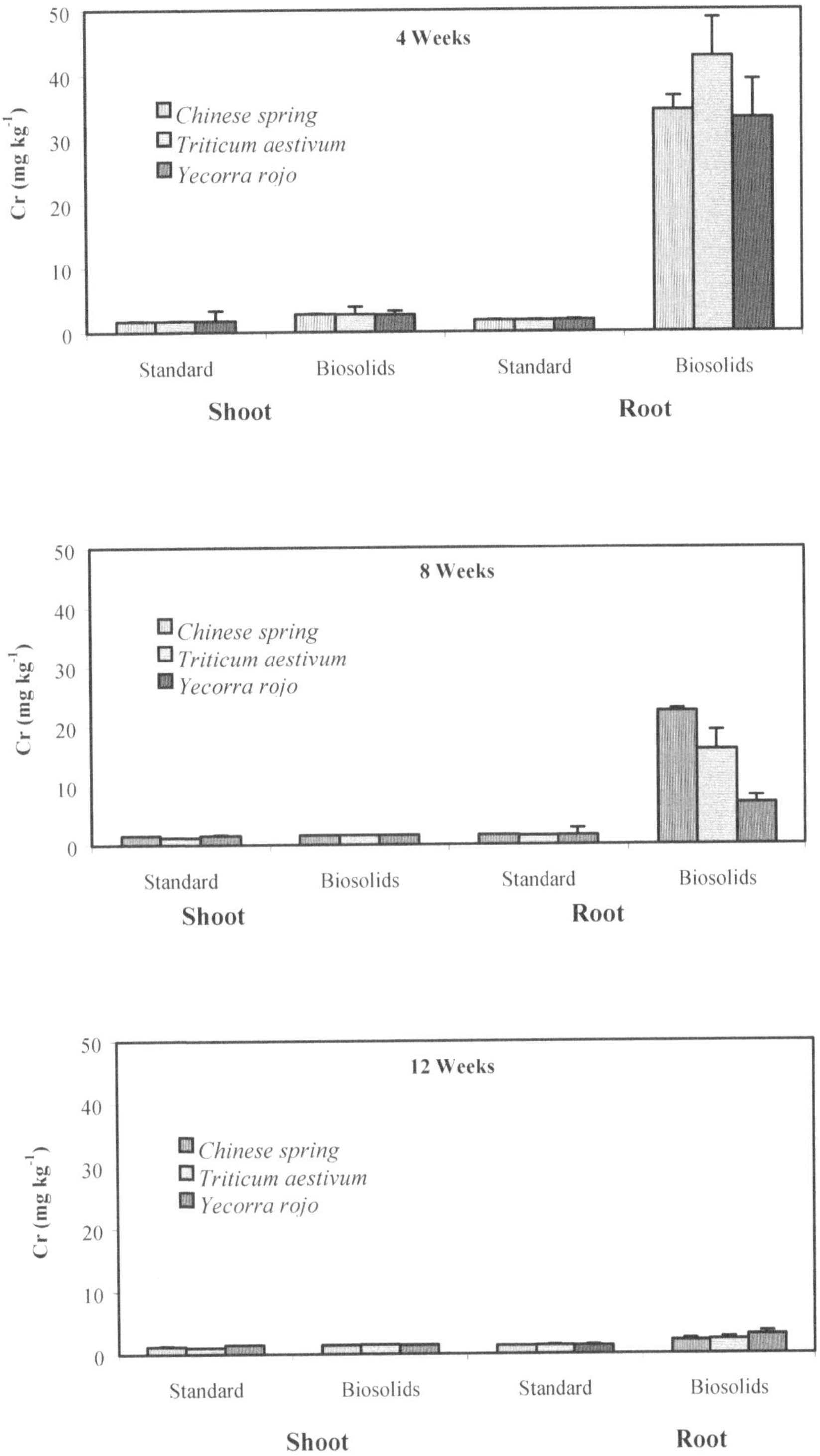

Figure 5-8. Cr Concentrations of Three Wheat Varieties Grown on Standard Nutrient Solution- and MWRDGC 1975 Biosolids-Treated Media.

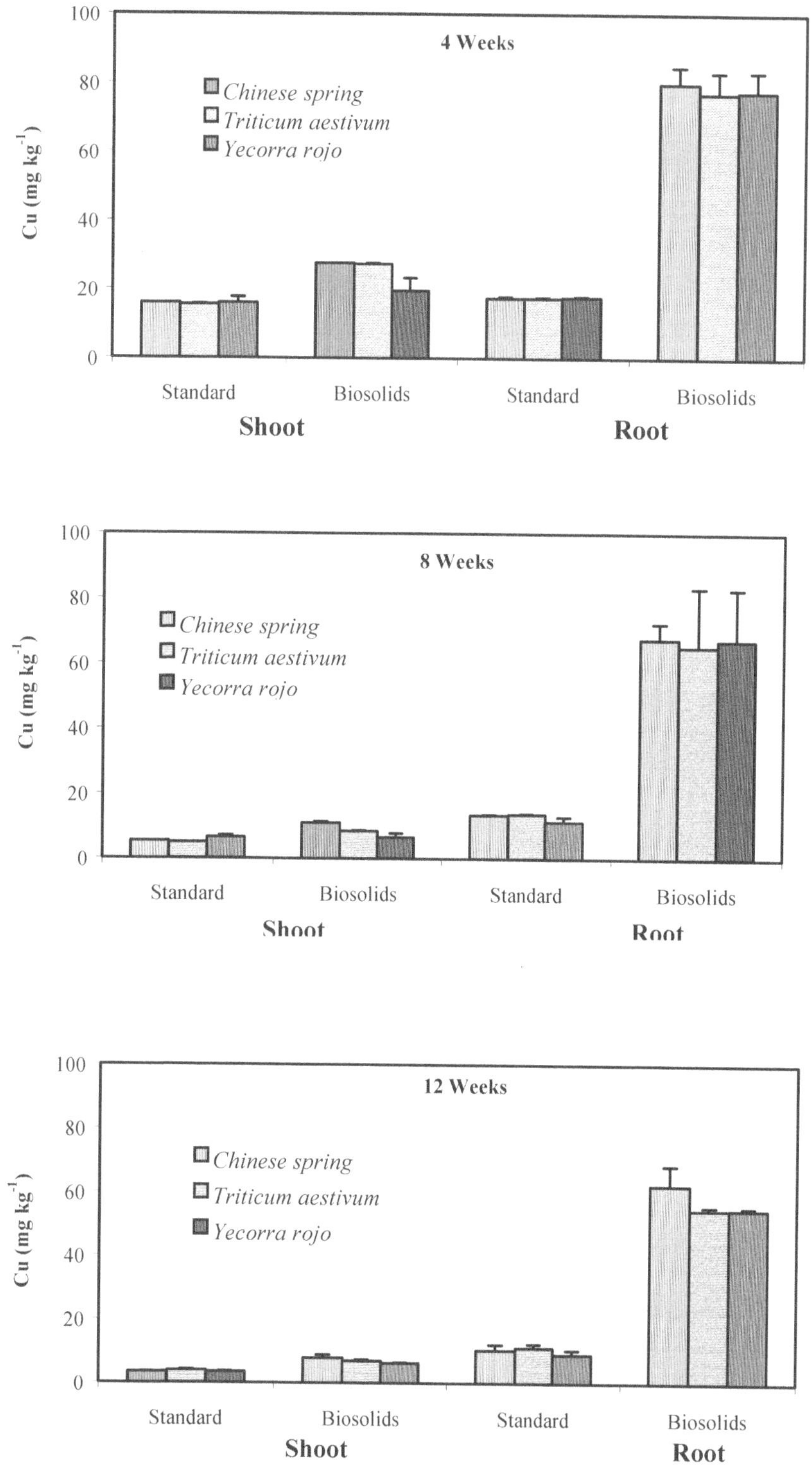

Figure 5-9. Cu Concentrations of Three Wheat Varieties Grown on Standard Nutrient Solution- and MWRDGC 1975 Biosolids-Treated Media.

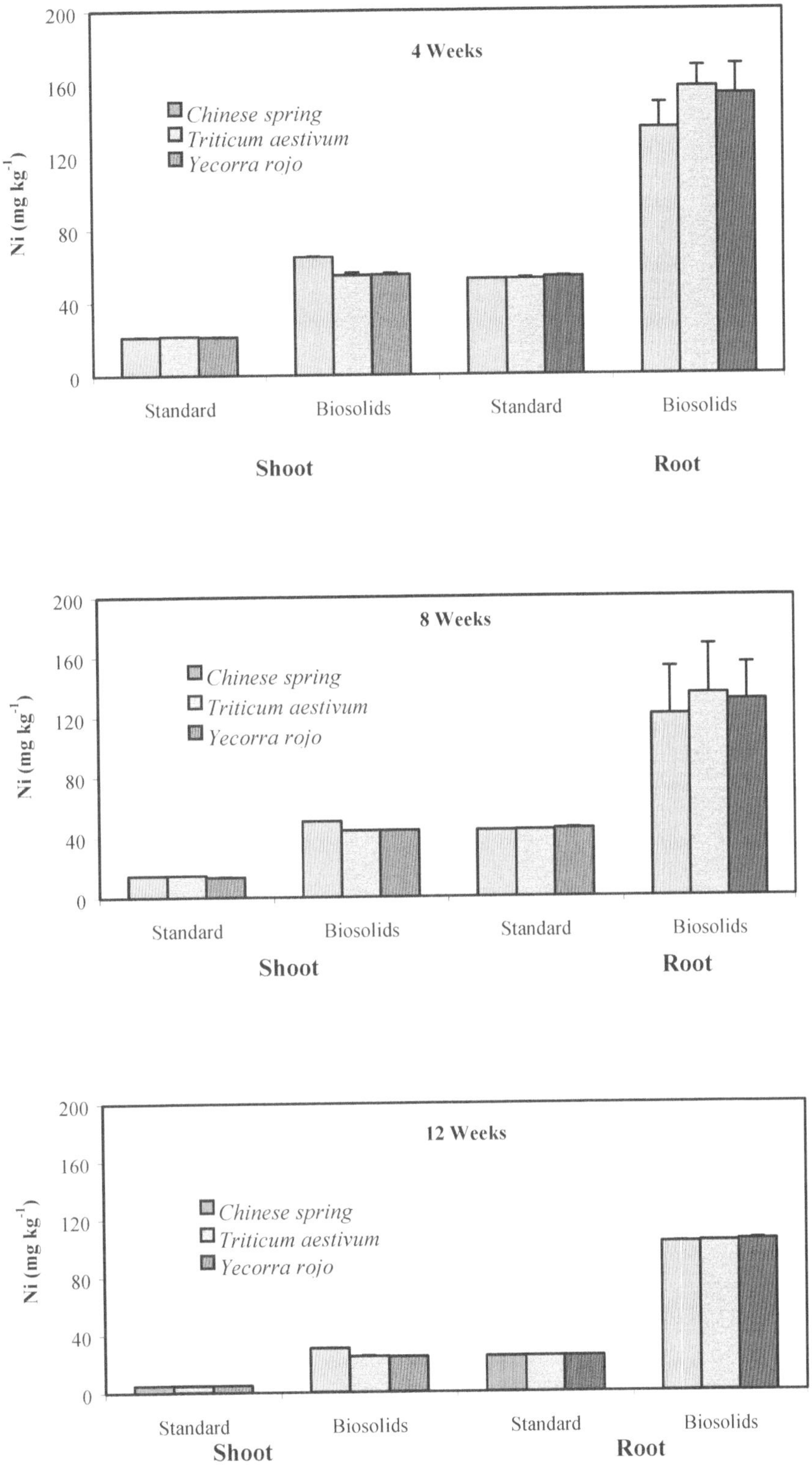

Figure 5-10. Ni Concentrations of Three Wheat Varieties Grown on Standard Nutrient Solution- and MWRDGC 1975 Biosolids-Treated Media.

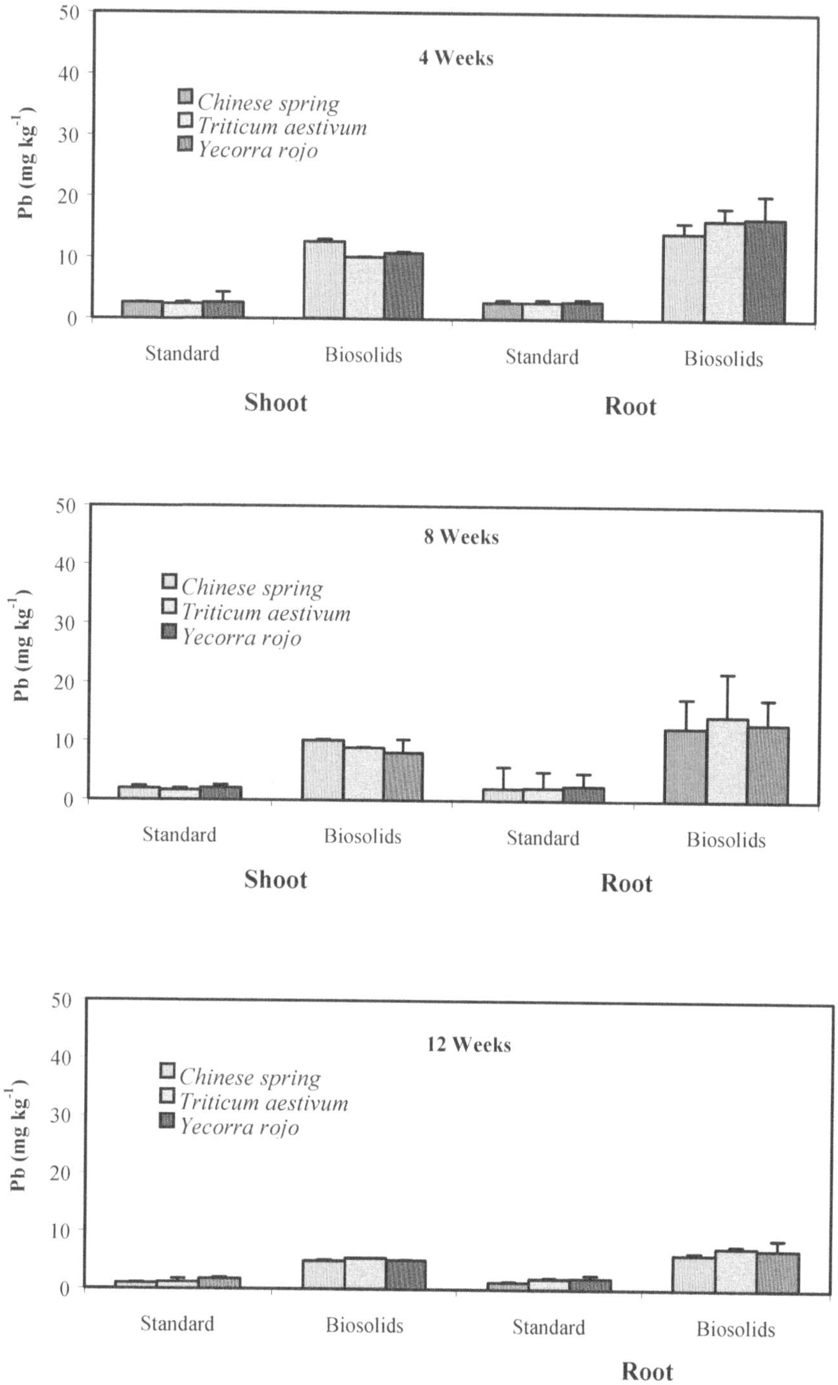

Figure 5-11. Pb Concentrations of Three Wheat Varieties Grown on Standard Nutrient Solution- and MWRDGC 1975 Biosolids-Treated Media.

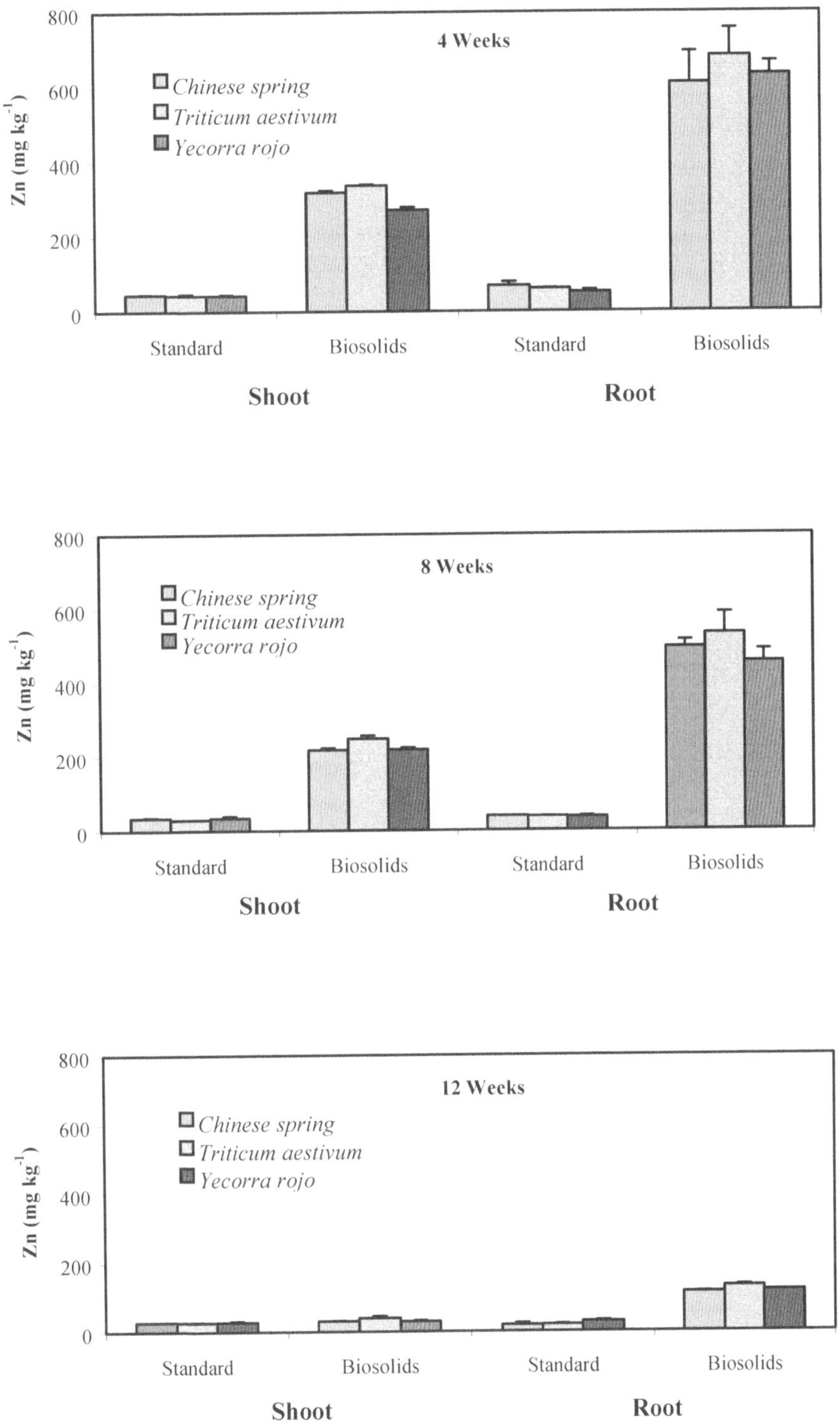

Figure 5-12. Zn Concentrations of Three Wheat Varieties Grown on Standard Nutrient Solution- and MWRDGC 1975 Biosolids-Treated Media.

Table 5-5. Comparison of Metal Concentrations in Three Cultivars of Wheat Grown on Nutrient Solution Irrigated Sand Media.

Element	Plant Part	Variety	Average Concentration (µg g^{-1})	
			0–4 Weeks	4–12 Weeks
Cd	Shoot	Chinese spring[a]	2.53	0.56
		Durum[b]	2.55	0.75
		Yecora Rojo[c]	2.56	1.01
	Root	Chinese spring	2.63	0.75
		Durum	2.64	0.78
		Yecora Rojo	2.84	0.72
Cr	Shoot	Chinese spring	1.75	1.54
		Durum	1.73	1.15
		Yecora Rojo	1.75	1.52
	Root	Chinese spring	1.77	1.45
		Durum	1.76	1.44
		Yecora Rojo	1.81	1.44
Cu	Shoot	Chinese spring	16.0	4.44
		Durum	15.5	4.46
		Yecora Rojo	16.0	5.07
	Root	Chinese spring	17.4	12.0
		Durum	17.3	12.5
		Yecora Rojo	17.7	10.3
Ni	Shoot	Chinese spring	21.1	9.82
		Durum	21.4	9.72
		Yecora Rojo	21.2	9.20
	Root	Chinese spring	52.5	34.4
		Durum	52.6	34.6
		Yecora Rojo	53.7	35.1
Pb	Shoot	Chinese spring	2.62	1.49
		Durum	2.41	1.41
		Yecora Rojo	2.63	1.93
	Root	Chinese spring	2.69	1.67
		Durum	2.72	2.03
		Yecora Rojo	2.87	2.23
Zn	Shoot	Chinese spring	45.7	31.5
		Durum	43.4	28.7
		Yecora Rojo	42.9	31.3
	Root	Chinese spring	68.5	28.5
		Durum	60.3	28.5
		Yecora Rojo	51.0	32.7

[a]*Tricutum turgidum* cv. Chinese spring; [b]*Tricutum aestivum L.* cv. Durum; [c]*Tricutum turgidum* cv. Yecora Rojo

Table 5-6. Comparison of Metal Concentrations in Three Cultivars of Wheat Grown on MWRDGC 1975 Biosolids-Treated Sand Media.

Element	Plant Part	Variety	Average Concentration (μg g^{-1})	
			0–4 Weeks	4–12 Weeks
Cd	Shoot	Chinese spring[a]	24.6	6.06
		Durum[b]	29.8	8.75
		Yecora Rojo[c]	24.4	8.47
	Root	Chinese spring	41.8	18.0
		Durum	39.3	12.9
		Yecora Rojo	26.1	15.8
Cr	Shoot	Chinese spring	2.77	1.51
		Durum	2.70	1.56
		Yecora Rojo	2.69	1.52
	Root	Chinese spring	34.5	12.2
		Durum	42.7	9.11
		Yecora Rojo	33.2	4.99
Cu	Shoot	Chinese spring	27.7	9.50
		Durum	27.4	7.73
		Yecora Rojo	20.0	6.41
	Root	Chinese spring	79.9	64.8
		Durum	77.1	59.9
		Yecora Rojo	77.5	60.9
Ni	Shoot	Chinese spring	64.9	40.5
		Durum	54.6	34.5
		Yecora Rojo	54.9	34.3
	Root	Chinese spring	135	112
		Durum	158	120
		Yecora Rojo	154	118
Pb	Shoot	Chinese spring	12.6	7.49
		Durum	10.2	7.13
		Yecora Rojo	10.8	6.57
	Root	Chinese spring	14.2	9.24
		Durum	16.2	10.9
		Yecora Rojo	16.6	10.0
Zn	Shoot	Chinese spring	319	124
		Durum	338	144
		Yecora Rojo	271	125
	Root	Chinese spring	610	304
		Durum	682	342
		Yecora Rojo	633	286

[a]*Tricutum turgidum* cv. Chinese spring; [b]*Tricutum aestivum L.* cv. Durum; [c] *Tricutum turgidum* cv. Yecora Rojo

When biomass production was taken into account, the actual metal uptake of the three wheat varieties could vary by a factor of two to three times (Table 5-7). Because of the smaller biomass, the amounts of biosolids-borne metals absorbed by wheat plants were considerably lower than that of corn grown for the same period of time (Table 5-1). In general, Chinese spring accumulated greater amounts of metals than the other two varieties, reflecting the difference in their respective biomass yields. The higher capacity of metal absorption by Chinese spring over the other two varieties might be due to its longer vegetative growth period. While Yecora Rojo and Durum reached maturity at 10–12 weeks, Chinese spring remained in the vegetative stage at the end of the 12th week.

Table 5-7. Percentage of Total Added Metals Absorbed by Shoot Tissues of Three Wheat VarietiesGrown on Biosolids-Treated Media for 12 Weeks.

Wheat Variety	Biosolids-Borne Metal Absorbed (%)					
	Cd	Cr	Cu	Ni	Pb	Zn
Chinese spring[a]	2.8	0.1	1.5	1.2	0.6	0.8
Durum[b]	3.5	0.1	0.8	0.6	0.4	0.7
Yecora Rojo[c]	1.0	0.04	0.5	0.4	<0.1	0.3

[a]*Tricutum turgidum* cv. Chinese spring; [b]*Tricutum aestivum L.* cv. Durum; [c]*Tricutum turgidum* cv. Yecora Rojo

The rates of metal uptake, measured in micrograms (μg) of metal accumulated per unit biomass per growing period, for the three wheat varieties varied with the development of the plants (Tables 5-8 and 5-9). For all of the elements, the uptake rates were higher at the earlier stage of growth and rapidly decreased over time. The rates of Cr and Pb uptake by plants were restricted by the availability of these two elements and were significantly lower than those of Zn, Cd, Cu, and Ni. The uptake by shoots and roots followed essentially the same pattern. The metal uptake rates among the three cultivars of wheat were similar.

Table 5-8. Rate of Metal Uptake by Shoots of Three Varieties of Wheat Grown on Standard Nutrient Solution Sand Media.

Growing Stage[a]	Wheat Variety	Metal Uptake Rate ($\mu g\ g^{-1}\ 4\ wk^{-1}$)					
		Cd	Cr	Cu	Ni	Pb	Zn
Early	Chinese spring[b]	12	2	14	32	6	160
	Durum[c]	15	2	14	27	5	169
	Yecora Rojo[d]	12	2	10	27	6	136
Rapid Growing	Chinese spring	4	1	4	23	5	96
	Durum	6	1	3	21	4	116
	Yecora Rojo	7	1	2	21	4	104

[a]Early and rapid growing denote 0–4 weeks and 4–12 weeks of the growing periods, respectively.
[b]*Tricutum turgidum* cv. Chinese spring; [c]*Tricutum aestivum L.* cv. Durum; [d]*Tricutum turgidum* cv. Yecora Rojo

Table 5-9. Rate of Metal Uptake by Shoots of Three Varieties of Wheat Grown on Biosolids-treated Sand Media.

Growing Stage[a]	Wheat Variety	Metal Uptake Rate ($\mu g\ g^{-1}\ 4\ wk^{-1}$)					
		Cd	Cr	Cu	Ni	Pb	Zn
Early	Chinese spring[b]	21	17	40	68	7	305
	Durum[c]	20	21	39	79	8	341
	Yecora Rojo[d]	18	17	39	77	8	316
Rapid Growing	Chinese spring	9	10	32	59	6	232
	Durum	5	2	30	62	7	227
	Yecora Rojo	7	15	26	49	8	100

[a]Early and rapid growing denote 0–4 weeks and 4–12 weeks of the growing periods, respectively.
[b]*Tricutum turgidum* cv. Chinese spring; [c]*Tricutum aestivum L.* cv. Durum; [d]*Tricutum turgidum* cv. Yecora Rojo

5.3 Metal Uptake by Six Plant Species Grown on Sand Media Treated with Nutrient Solution, MWRDGC 1975 Biosolids, and Phosphorus Rock

In this experiment, six plant species were grown for eight weeks on nutrient solution irrigated sand media, sand media treated with 20 Mg ha^{-1} of MWRDGC 1975 biosolids, and sand media treated with 100 kg ha^{-1} of phosphorus rock. The metal absorption by Durum wheat, Sudan grass, corn, canola (rape), chickpea, and Swiss chard that were grown on MWRDGC 1975 biosolids-treated media followed the same general patterns of metal uptake as plants described in previous sections (Figures 5-13 to 5-18). In this case, a phosphorus rock treatment was included as a reference for plant responses to a relatively unavailable phosphorus nutrient. The metal concentrations of plant tissue for the respective treatments were for treatment: biosolids > standard plant nutrient = phosphorus rock, and for plant parts: root > shoot.

For each metal element, the concentrations for plants grown on biosolids-treated media were significantly different between the plant species. For Cd, which was an element readily absorbed by plants, the average concentrations among the six species varied from 1.2 to 1.8 mg kg^{-1} for those grown on standard nutrient solution treated sand media and were not significantly different between plant species. The concentrations ranged from 14–29 mg kg^{-1} for plants grown under biosolids treatments and were significantly different between the plant species tested. The Cd concentration of plant shoots under the phosphorus rock treatment was slightly higher than those from the standard plant nutrient solution treatment (1.4–4.9 mg kg^{-1}), possibly reflecting the Cd present in the phosphorus rock. Cr was not readily accumulated in the plant shoots and the differences of plant tissue concentrations between the treatments were considerably narrower (Figure 5-14). Cr concentrations in plant shoots were 0.9 to 1.0, 1.3 to 4.6, and 0.7 to 1.9 mg kg^{-1} for the standard plant nutrient solution, biosolids, and phosphorus rock treatments, respectively. However, the Cr concentrations in the plant roots were 1.2–1.7, 17–77, and 1.0–3.9 mg kg^{-1} for the standard plant nutrient, biosolids, and phosphorus rock treatments, respectively (Figure 5-14). The higher concentrations in the roots and lower concentrations in the shoots appeared to indicate that significant amounts of metal solubilized in the rhizosphere were absorbed by roots and were not translocated into the shoots.

Different plants species, however, accumulated different amounts of metals in the plant tissue (Table 5-10) and, except for Cr, the metal accumulation in the shoots was larger than that in the corresponding root tissue.

Table 5-10. Metal Uptake by Shoots and Roots of Different Plant Species Grown on Biosolids-Treated Media.

Plant Part	Species	Metal Uptake by Shoots (μg per pot)					
		Cd	Cu	Zn	Ni	Cr	Pb
Shoot	Durum wheat	0.26	0.18	2.51	0.74	0.03	0.12
	Sudan grass	0.37	0.22	5.77	1.19	0.03	0.13
	Corn	0.38	0.21	3.98	0.73	0.02	0.19
	Canola (Rape)	0.21	0.22	3.37	0.33	0.06	0.18
	Chickpea	0.09	0.11	1.38	0.34	0.03	0.06
	Swiss chard	0.18	0.14	2.07	0.16	0.03	0.04
Root	Durum wheat	0.10	0.30	1.98	0.54	0.14	0.06
	Sudan grass	0.16	0.50	3.40	1.04	0.15	0.06
	Corn	0.17	0.30	2.42	1.22	0.22	0.11
	Canola (Rape)	0.05	0.20	0.76	0.23	0.17	0.07
	Chickpea	0.14	0.21	2.58	0.70	0.07	0.05
	Swiss chard	0.07	0.28	0.80	0.47	0.04	0.01

The amounts of trace elements accumulated in the plant shoot followed the general descending trends:

Cd: Sudan grass = Corn > Wheat > Canola (rape) = Swiss chard > Chickpea
Cr: Sudan grass = Corn = Wheat = Canola (rape) = Swiss chard = Chickpea
Cu: Sudan grass = Corn = Canola (rape) >Wheat > Swiss chard > Chickpea
Ni: Sudan grass > Corn = Wheat > Canola (rape) = Chickpea > Swiss chard
Pb: Canola (rape) = Corn > Sudan grass = Wheat > Chickpea > Swiss chard
Zn: Sudan grass > Corn > Canola (rape) > Wheat > Swiss chard > Chickpea

Greater percentages of biosolids-borne Cd, Ni, and Zn in the rhizosphere were absorbed by plants than were Cr and Pb (Table 5-11).

Table 5-11. Percent of the Total Metal Absorbed by Shoots and Roots of Different Plant Species Grown on Biosolids-Treated Media.

Plant Part	Species	Biosolids-Borne Metal Absorbed (%)					
		Cd	Cu	Zn	Ni	Cr	Pb
Shoot	Durum wheat	2.36	0.25	0.61	1.62	0.02	0.10
	Sudan grass	3.32	0.31	1.40	2.62	0.02	0.10
	Corn	3.38	0.30	0.96	1.61	0.02	0.15
	Canola (Rape)	1.90	0.30	0.82	0.72	0.04	0.14
	Chickpea	0.85	0.15	0.33	0.75	0.02	0.05
	Swiss chard	1.62	0.20	0.50	0.36	0.02	0.03
Root	Durum Wheat	0.90	0.41	0.48	1.19	0.09	0.05
	Sudan Grass	1.46	0.70	0.82	2.31	0.10	0.05
	Corn	1.50	0.41	0.59	2.69	0.15	0.08
	Canola (Rape)	0.44	0.28	0.18	0.50	0.12	0.05
	Chickpea	1.25	0.29	0.63	1.54	0.05	0.04
	Swiss Chard	0.59	0.39	0.19	1.04	0.03	0.01

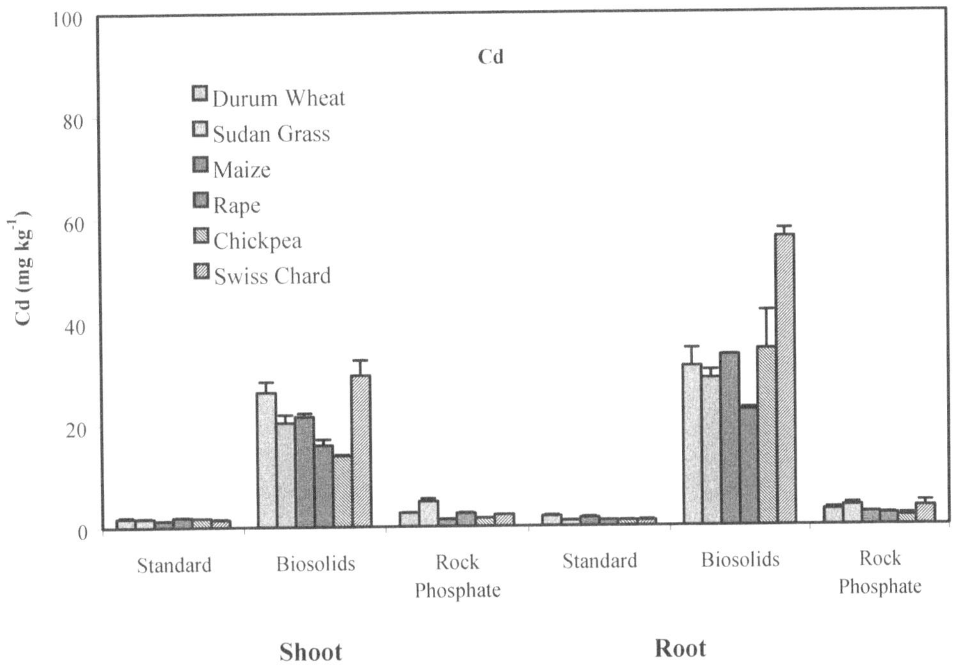

Figure 5-13. Cd Concentrations of Durum Wheat, Sudan Grass, Corn, Canola (Rape), Chickpea, and Swiss Chard Grown on Nutrient Solution, Biosolids-Treated and Phosphorus Rock-Treated Media.

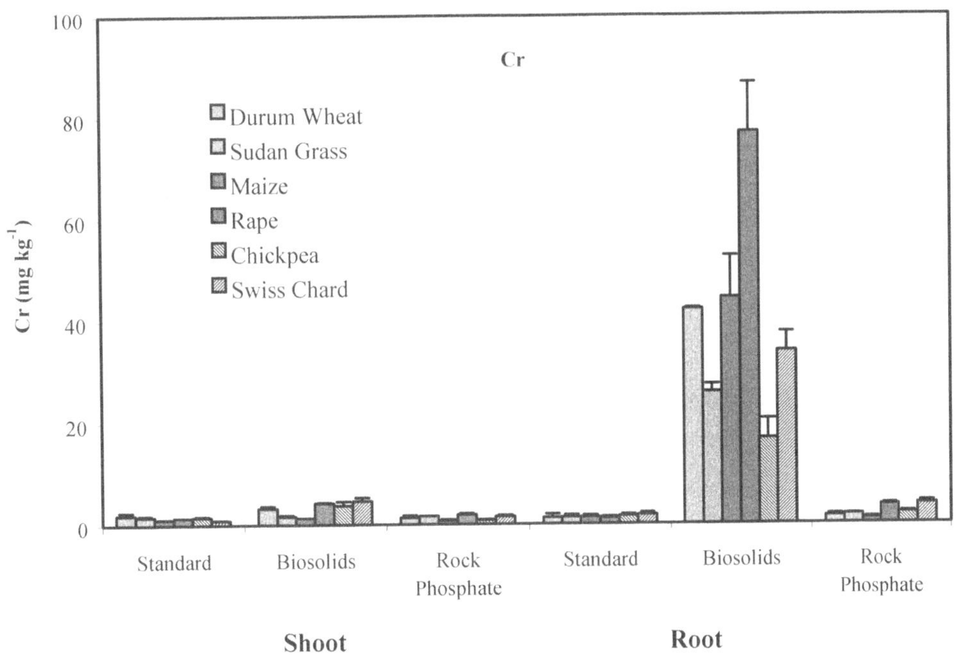

Figure 5-14. Cr Concentrations of Durum Wheat, Sudan Grass, Corn, Canola (Rape), Chickpea, and Swiss Chard Grown on Nutrient Solution, Biosolids-Treated and Phosphorus Rock-Treated Media.

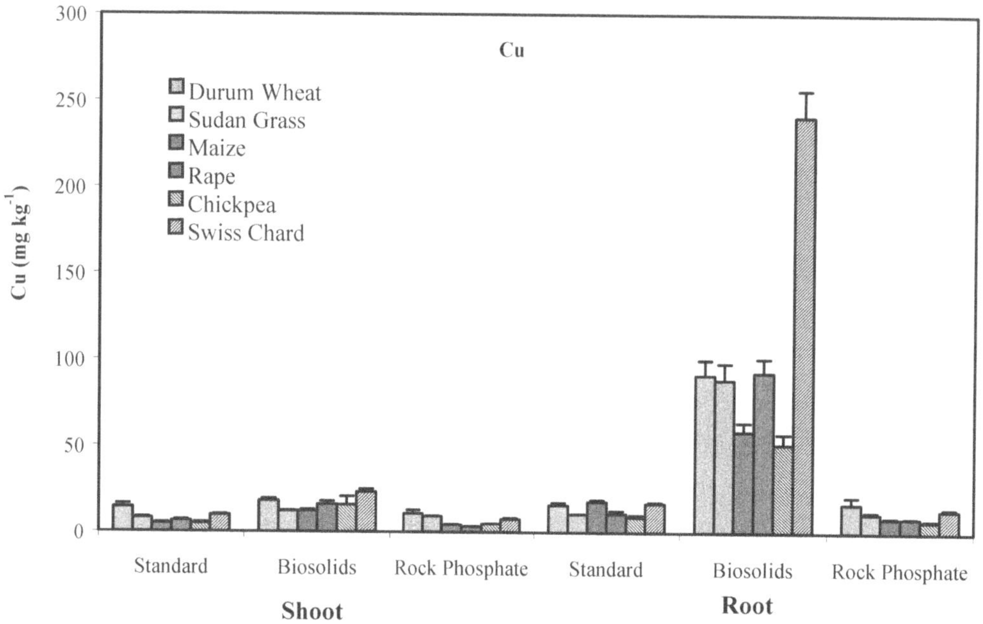

Figure 5-15. Cu Concentrations of Durum Wheat, Sudan Grass, Corn, Canola (Rape), Chickpea, and Swiss Chard Grown on Nutrient Solution, Biosolids-Treated and Phosphorus Rock-Treated Media.

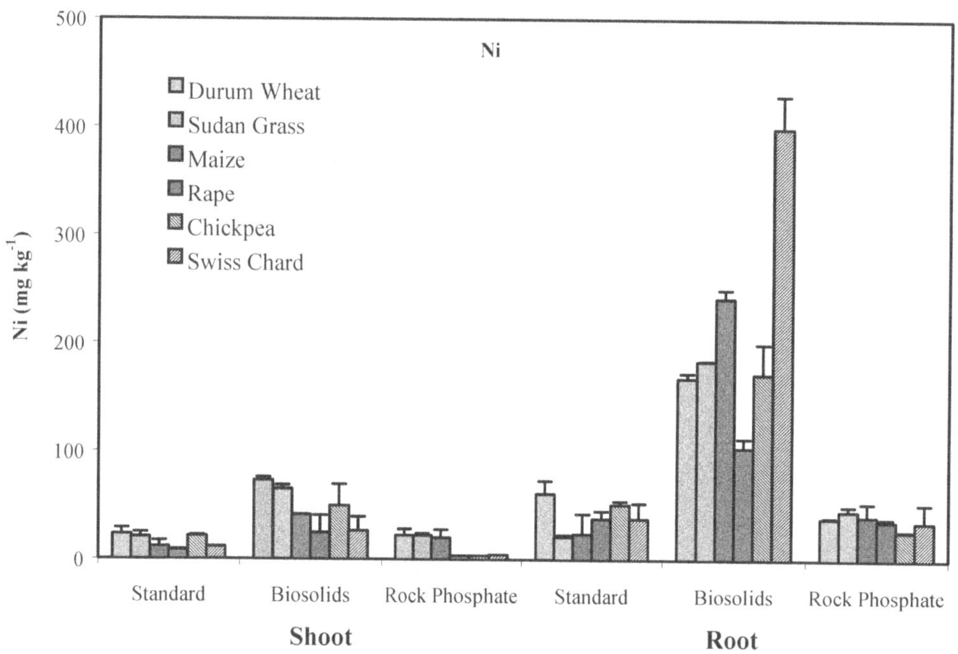

Figure 5-16. Ni Concentrations of Durum Wheat, Sudan Grass, Corn, Canola (Rape), Chickpea, and Swiss Chard Grown on Nutrient Solution, Biosolids-Treated and Phosphorus Rock-Treated Media.

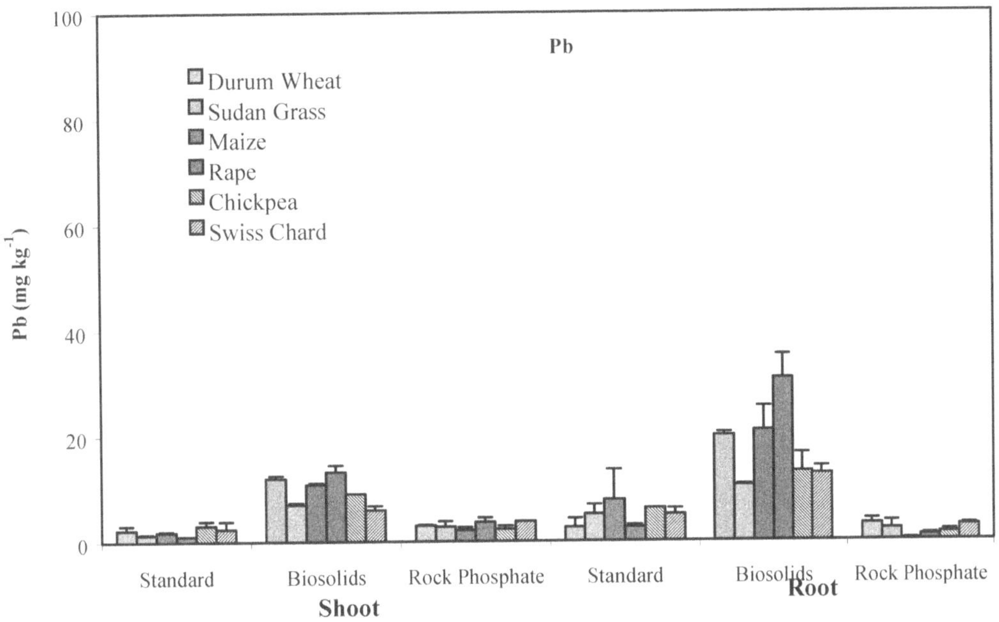

Figure 5-17. Pb Concentrations of Durum Wheat, Sudan Grass, Corn, Canola (Rape), Chickpea, and Swiss Chard Grown on Nutrient Solution, Biosolids-Treated and Phosphorus Rock-Treated Media.

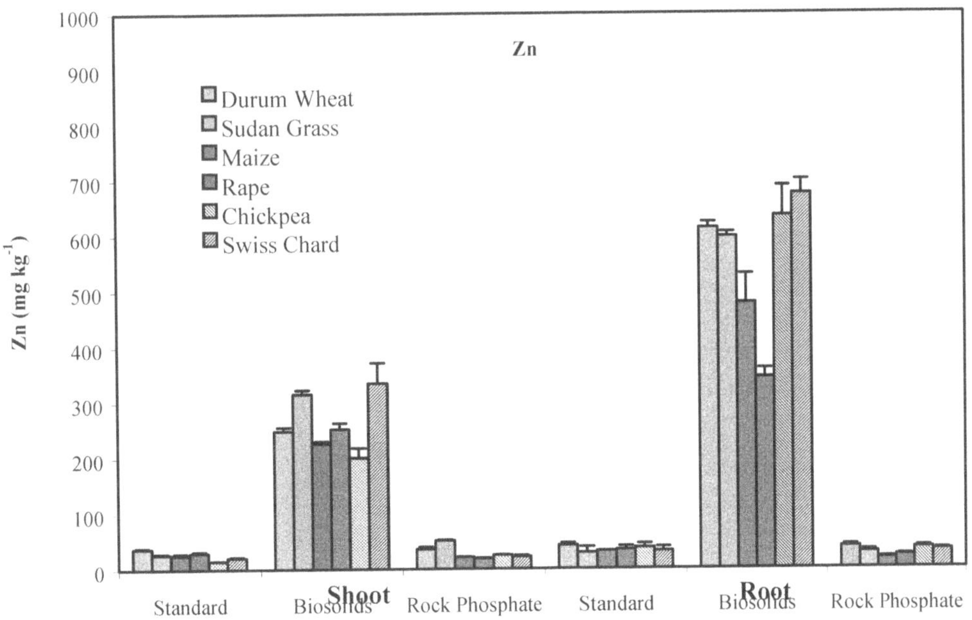

Figure 5-18. Zn Concentrations of Durum Wheat, Sudan Grass, Corn, Canola (Rape), Chickpea, and Swiss Chard Grown on Nutrient Solution, Biosolids-Treated and Phosphorus Rock-Treated Media.

The monocots (Durum wheat, Sudan grass, and corn) were not able to use the P in the phosphorus rock as effectively as the P from the nutrient solution or the biosolids; this consequently limited the biomass yield of plants grown in the phosphorus rock treatment (Table 5-12). On the other hand, the dicots (canola, chickpea, and Swiss chard) used the P in phosphorus rock effectively and the P content of the plant tissues receiving nutrient solution, biosolids, and phosphorus rock were comparable.

Table 5-12. Phosphorus Concentrations of Plants Grown on Standard Nutrient Solution, Biosolids-Treated, and Phosphorus Rock-Treated Media.

Plant	Treatment	Phosphorus ($\mu g\ g^{-1}$)	
		Shoot	Root
Chickpea	Nutrient Solution	1,260 c	922 c
	Biosolids	2,044 b	1,137 b
	Phosphorus Rock	2,580 a	1,513 a
Rape	Nutrient Solution	1,743 c	1,564 c
	Biosolids	2,189 b	2,203 b
	Phosphorus Rock	3,354 a	2,707 a
Swiss chard	Nutrient Solution	1,753 c	1,372 c
	Biosolids	2,141 a	3,261 a
	Phosphorus Rock	2,029 a	2,631 b
Corn	Nutrient Solution	1,198 a	707 b
	Biosolids	1,192 a	763 a
	Phosphorus Rock	638 b	556 c
Sudan grass	Nutrient Solution	1,088 b	676 b
	Biosolids	1,138 a	814 a
	Phosphorus Rock	756 c	553 c
Durum wheat	Nutrient Solution	1,744 b	956 b
	Biosolids	2,537 a	1,716 a
	Phosphorus Rock	783 c	638 c

Note: Values represent the mean of 3 replicates. Letters following the value indicate statistical differences. For each species and plant part, values followed by the same letter were not significantly different at $p < 0.05$.

The organic acids in root exudates are thought to be important for the solubilization of sparingly soluble P in phosphorus rock. The factors most likely to limit the P dissolution are insufficient pH buffering and/or P sorption capacity, and increase in Ca^{2+} activity in the soil solution around phosphorus rock particles. Phosphate rocks have been reported to be more effective in supplying P to dicots (i.e., white clover) than to monocots (i.e., wheat and barley) (Gregg et al., 1988). Dicots can acidify their rhizosphere more effectively when they fix nitrogen. Nitrogen enters the root as an uncharged molecule and an excess of cations over anions can be absorbed by the root to balance the negative charge on newly formed proteins and organic acids, which are products of carbon and nitrogen assimilation. The excess cation uptake is balanced by a net H^+ efflux into the rhizosphere (Nyatsanga and Pierre, 1973; Bolan et al., 1989). Some organic acids (i.e., citric and malic) are well known for their capacity to desorb phosphorus from sesquioxide surfaces by anion exchange (Parfitt, 1979). Both desorption and chelation processes are responsible for phosphate rock solubilization. The capacity of organic acids in root exudates to solubilize phosphorus rock also depends on their concentration per unit volume of soil. Thus, it can be anticipated that with increasing density of root hairs or of lateral roots, the likelihood of organic acids mobilizing phosphorus rock becomes higher (Gardner et al., 1983a).

Later in this report, we report that the composition of root-borne dicarboxylic organic acids (i.e., succinic, tartaric, and oxalic) varied between dicots and monocots and the total amount and production rate of organic acids from dicots were higher than those from monocots.

5.4 Plant Uptake of Metals from Biosolids of Five Sources

The uptake of metals from biosolids-treated media reported in the previous sections was conducted with the MWRDGC 1975 biosolids that had considerably higher metals contents than biosolids currently produced by POTWs across the nation. In this experiment, we compared the metal uptake of corn grown on the sand media amended with the MWRDGC 1975 biosolids with those grown on sand media amended with comparable amounts of biosolids collected from four POTWs in 1999 (i.e., MWRDGC 1999, Milorganite, Los Angeles, and Riverside). The concentrations of metals in the 1999-collected biosolids were significantly lower than the metal contents of MWRDGC 1975 (Table 2-2). Corn responded to the biosolids treatments and produced significantly greater biomass than that on the standard nutrient solution treatment (Figure 4-5).

The amounts of metals accumulated in shoots of corn are presented in Figures 5-19 through 5-24. For the convenience in comparing data presented in the figures, the scales of the Y-axis for the figures were kept the same. Based on the metals concentrations in the plant tissue, the metal uptake was calculated at weeks 4, 8, and 12 of the growing (Figures 5-19 to 5-24). Greater percentages of the total biosolids-borne Cd and Zn in the rhizosphere were available to plants than the other elements (Table 5-13).

Table 5-13. Percent of the Total Metal Absorbed by Corn Grown for 12 Weeks on Different Biosolids-Treated Media.

Source	Biosolids-borne Metal Absorbed (%)					
	Cd	Cr	Cu	Ni	Pb	Zn
MWRDGC 1999	5.8	1.1	1.7	7.0	0.2	5.5
Riverside	3.9	0.8	3.4	2.8	4.5	6.4
Los Angeles	3.6	0.8	2.2	2.2	1.6	6.0
MWRDGC 1975	16.3	0.1	1.7	0.8	0.1	10.9
Milorganite	7.9	0.3	5.2	1.7	2.1	6.2

The metal accumulation in the plant tissues and their relationships to the metal concentrations in the biosolids are summarized in Table 5-14.

Table 5-14. Relative Metal Accumulation in Tissues of Plants Grown on Biosolids with Varying Metals Concentration.

Element	Metal Concentration in Biosolids	Metal Accumulation in Plants
Cd	MWRDGC 1975 >>> LA >MWRDGC 1999 ≅ Milorganite > Riverside	MWRDGC 1975 >>> LA > MWRDGC 1999 ≅ Milorganite > Riverside
Cr	MWRDGC 1975 >>> Milorganite > MWRDGC 1999 > LA > Riverside	MWRDGC 1975 ≅ Milorganite ≅ MWRDGC 1999 ≅ LA ≅ Riverside
Cu	LA ≅ MWRDGC 1975 > Riverside > MWRDGC 1999 > Milorganite	Riverside > LA > MWRDGC 1975 > Milorganite > MWRDGC 1999
Ni	MWRDGC 1975 > LA > MWRDGC 1999 ≅ Milorganite > Riverside	MWRDGC 1975 ≅ MWRDGC 1999 ≅ Milorganite > LA > Riverside
Pb	MWRDGC 1975 >>> MWRDGC 1999 > Milorganite ≅ Riverside > LA	Inconsistent data for the evaluation
Zn	MWRDGC 1975 > LA > Riverside ≅ MWRDGC 1999 > Milorganite	MWRDGC 1975 > LA ≅ Riverside ≅ Milorganite ≅ MWRDGC 1999

Except for Cd, the plant uptake of metals by plants was not directly related to the metal concentrations in the biosolids. Other factors appear to have affected a metal's dissolution in the rhizosphere.

For Cd, Cu, Ni, and Zn, the uptake rates by plants grown on MWRDGC 1975 biosolids-treated media were considerably higher than respective rates for plants grown on media treated with other sources of biosolids (Table 5-15).

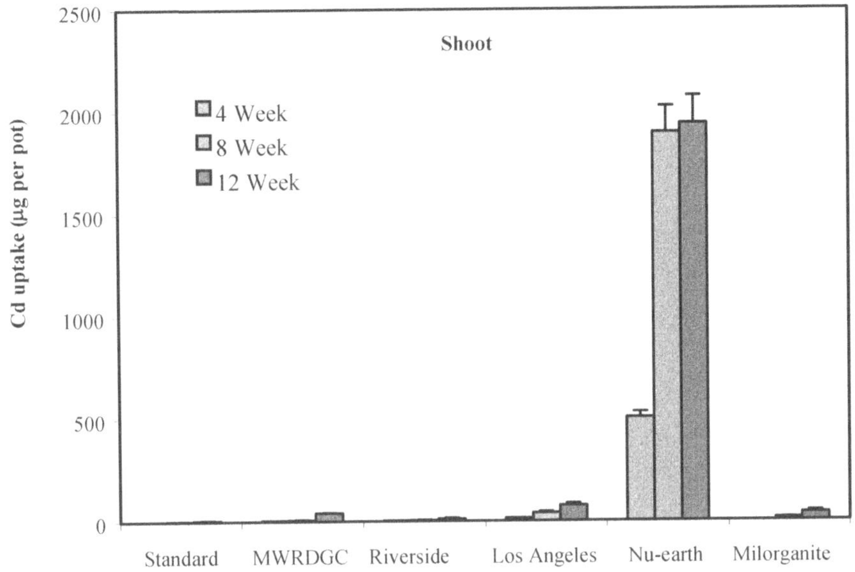

Figure 5-19. Cd Uptake by Corn Grown on Media Treated with Biosolids from Five Sources.

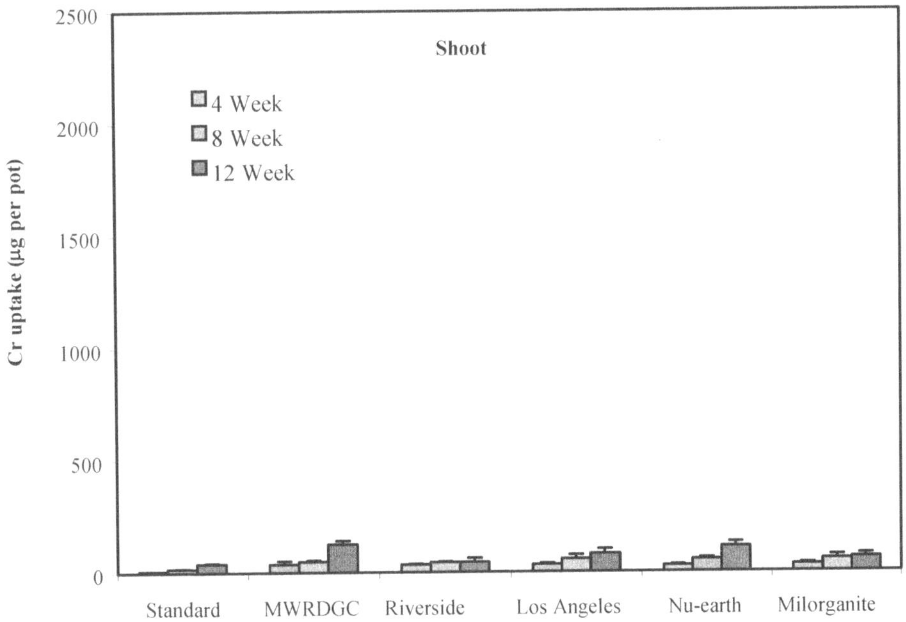

Figure 5-20. Cr Uptake by Corn Grown on Media Treated with Biosolids from Five Sources.

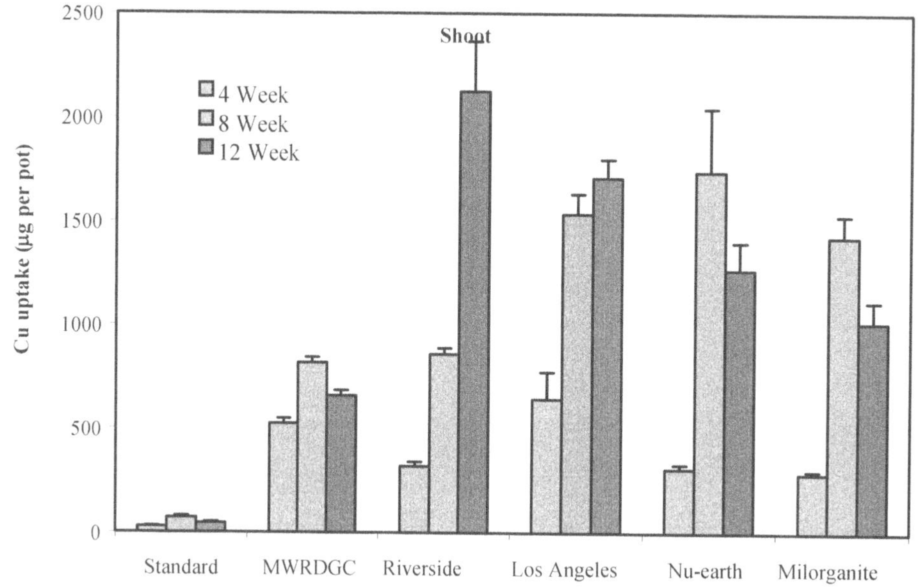

Figure 5-21. Cu Uptake by Corn Grown on Media Treated with Biosolids from Five Sources.

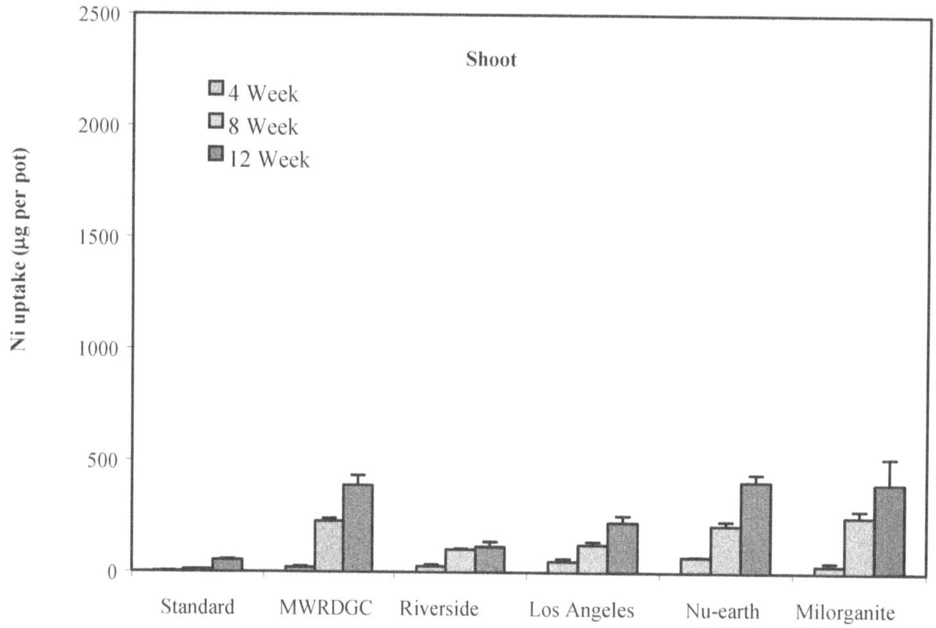

Figure 5-22. Ni Uptake by Corn Grown on Media Treated with Biosolids from Five Sources.

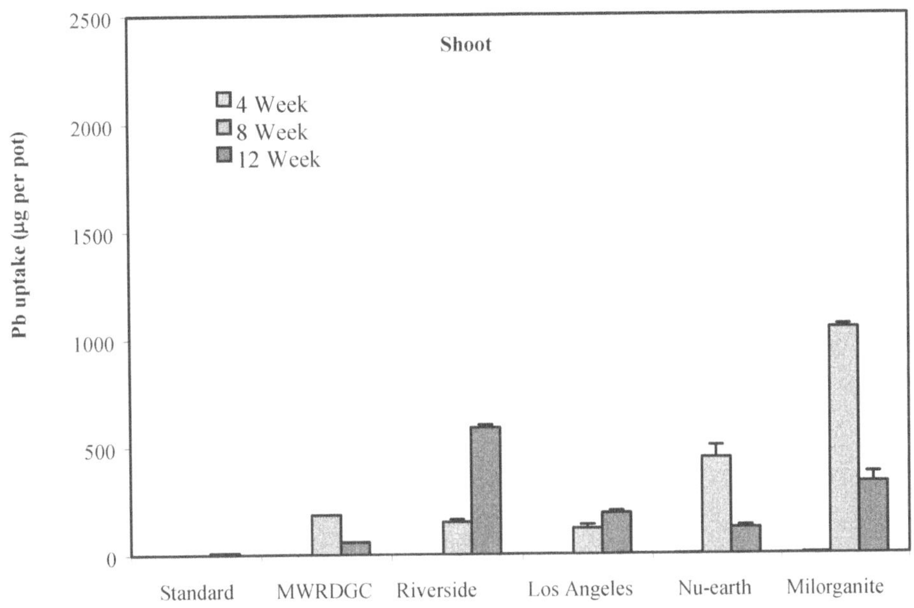

Figure 5-23. Pb Uptake by Corn Grown on Media Treated with Biosolids from Five Sources.

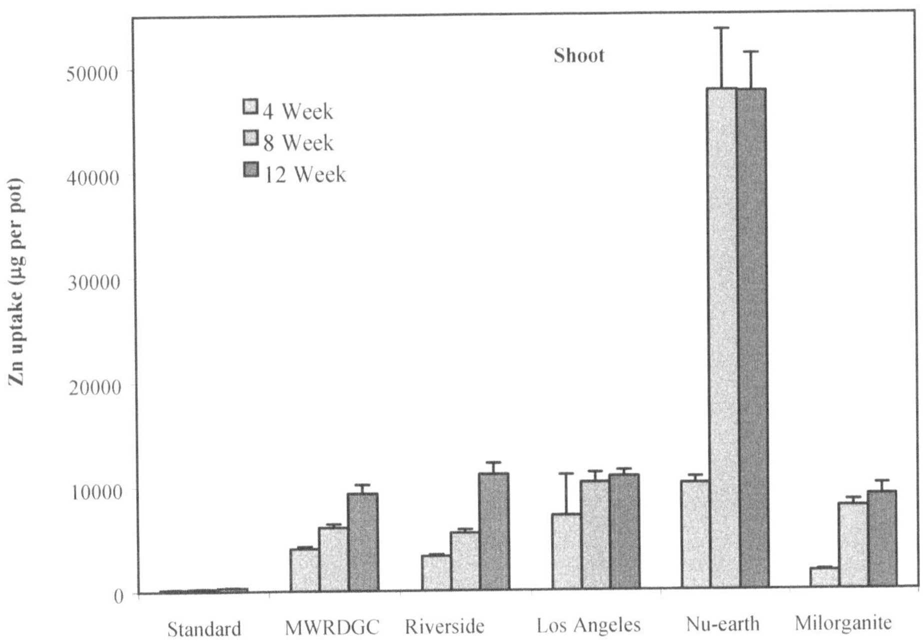

Figure 5-24. Zn Uptake by Corn Grown on Media Treated with Biosolids from Five Sources.

Table 5-15. Rate of Metals Uptake by Corn Shoots from Five Sources of Biosolids.

Element	Time (Week)	Uptake Rate (μg g^{-1} 4 wk^{-1})					
		Control	MWRDGC 1999	Riverside	Los Angeles	MWRDGC 1975	Milorganite
Cd	4	0.01	0.07	0.05	0.23	19	n.d.[a]
	8	0.08	0.03	0.03	0.18	23	0.08
	12	0.15	0.15	0.04	0.24	12	0.16
Cr	4	0.7	0.6	0.6	0.6	1.1	0.8
	8	0.9	0.2	0.2	0.3	0.7	0.4
	12	1.3	0.5	0.1	0.3	0.7	0.3
Cu	4	3	6	5	11	12	7
	8	4	4	4	7	21	9
	12	2	2	7	6	8	4
Ni	4	0.5	0.3	0.4	0.8	2.7	0.8
	8	0.6	1.1	0.5	0.6	2.6	1.6
	12	1.8	1.4	0.4	0.7	2.6	1.7
Pb	4	n.d.	n.d.	n.d.	n.d.	0.1	0.1
	8	0.03	0.9	0.8	0.6	5.5	6.7
	12	0.3	0.2	1.8	0.6	0.8	1.4
Zn	4	10	63	62	60	261	39
	8	14	29	28	48	587	51
	12	11	34	35	35	298	38

[a]n.d., not detected

5.5 Effect of Biosolids Particle Size on the Plant Absorption of Metals

Many factors may affect plant uptake of trace elements in soils; among them are pH, oxidation/reduction potential (Eh), cation exchange capacity (CEC), water regime, clay content, organic content, nutrient balance, and concentration of other trace elements (Kabata-Pendias, 2001). Generally, heavy metals are present in biosolids in sparingly soluble chemical forms. Upon entering the soil, the biosolids-borne metals remained in the solid phases. Therefore, the absorption of biosolids-borne metals by plants must take place at the root-metal solid interface where metals may dissolve into soil solution thereby contacting the root or roots may come into direct contact with the metal-containing solid particles. If this is the case, the physical attributes of biosolids, especially particle size, will influence the plant uptake of metals as smaller particles would provide greater interface area for the reaction to take place. The morphology and structural characteristics of biosolids particles generally are dictated by the chemical additives employed in dewatering and the follow up processing options, such as heat treatment, and composting.

5.5.1 Specific Surface Area and Heavy Metal Contents of Size-Classified Biosolids

The specific surface areas of the <0.833, 0.833-2.00 and 2.00-4.00 mm fractions of the MWRDGC 1975 biosolids ranged from 5.21 to 4.73 $m^2\ g^{-1}$ and were not significantly different from one another. The average specific surface area for the MWRDGC 1975 biosolids was 5 ± 0.25 $m^2\ g^{-1}$. For the Los Angeles biosolids, the specific surface areas were 1.39, 0.51, and 0.44 $m^2\ g^{-1}$ for the <0.833, 0.833-2.00 and 2.00-4.00 mm size fractions, respectively and they were 3.7, 9.9 and 10.8 fold lesser than that of the MWRDGC 1975 biosolids in comparable size ranges (Figure 5-25). The differences in specific surface area for these two biosolids may be attributed to the particles' morphological features. Images produced by the scanning electron microscopy showed that particles of MWRDGC 1975 biosolids were aggregated, porous, and spongy and the surface was rough with both cavities and extrusions (Figure 5-26). In addition to the external surface, these particles possessed substantial internal passages and surface areas. The biosolids of Los Angeles were discrete, granular, solid, and blocky and the particle surface was flat and smooth with few exterior features (Figure 5-26). It was apparent that the enhanced specific surface area in the MWRDGC 1975 biosolids was due to its porous particle structure that created interconnected internal passages and internal surface area that would facilitate the interfacial transfer of nutrients and metals from the biosolids to soil solution and/or the root surface.

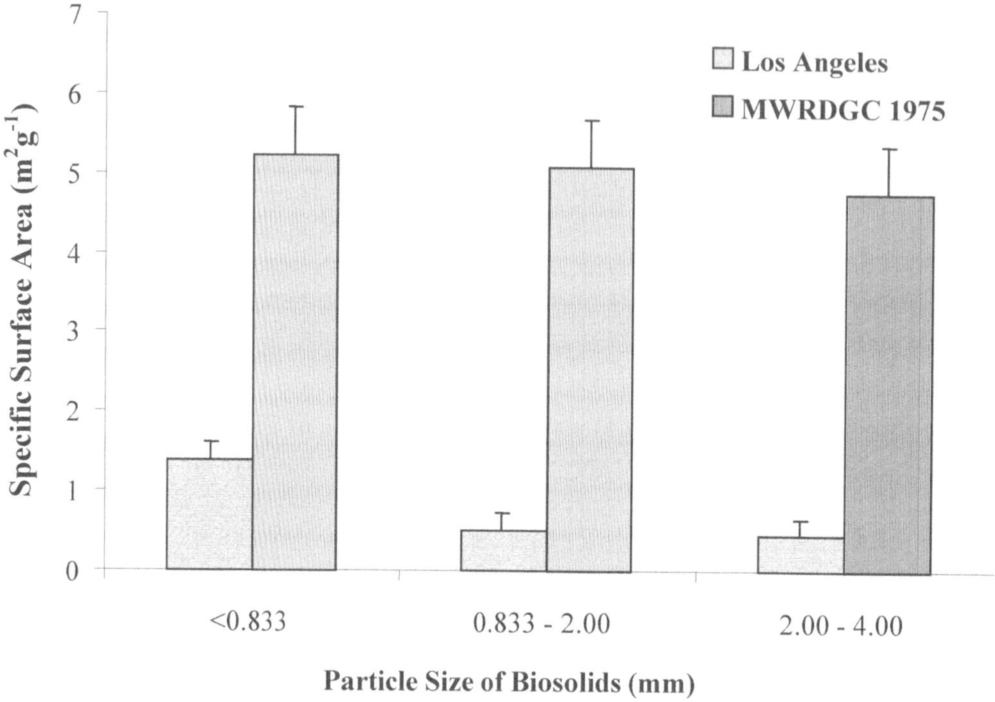

Figure 5-25. Specific Surface Area of Particle Size Classified Los Angeles and MWRDGC 1975 Biosolids. (Error bars indicate the LSD at $p < 0.05$.)

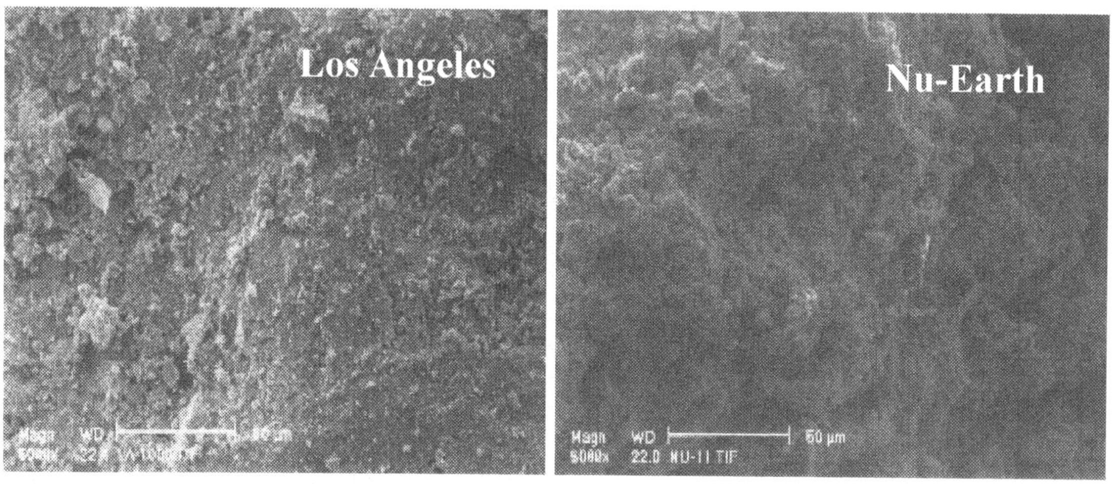

Figure 5-26. Scanning Electron Microscope Images of Particle Size-Classified Los Angeles and MWRDGC 1975 (Nu-earth) Biosolids.

For both biosolids, the metal concentrations were not significantly different between respective size-classified fractions (Tables 5-16 and 5-17). The nitrogen contents of the size-classified fractions followed the same patterns and the average N contents were 4.4% and 0.8% for the Los Angeles and MWRDGC 1975 biosolids, respectively. In this manner, the inputs of biosolids mass, nitrogen, and metals were nearly identical for each particle size treatment. The treatment differences lie in the number of particles and the surface areas created by respective biosolids particles. The heavy metal contents in MWRDGC 1975 biosolids, however, were considerably higher than those in the Los Angeles biosolids. The nitrogen contents of these two biosolids were also different.

Table 5-16. Metal Concentrations in Particle Size-Classified Los Angeles Biosolids.

Particle size (mm)	Metal Concentrations (mg kg^{-1})					
	Ni	Zn	Cu	Cr	Cd	Pb
<0.83	73	1239	760	86	18	41
0.83-2.00	73	1259	762	85	21	42
2.00-4.00	78	1260	780	88	22	45
4.00-6.35	77	1270	778	83	19	44
6.35-9.91	72	1198	750	82	18	39

Table 5-17. Metal Concentrations in Particle Size Classified MWRDGC 1975 Biosolids.

Particle size (mm)	Metal Concentrations (mg kg^{-1})					
	Ni	Zn	Cu	Cr	Cd	Pb
<0.83	510	5877	1151	2294	165	848
0.83-2.00	502	5789	1146	2184	162	833
2.00-4.00	511	5905	1162	2236	166	868
4.00-6.35	513	6014	1191	2225	168	877
6.35-9.91	486	5601	1147	2308	160	817

5.5.2 Biomass Accumulation and Plant Uptake of Metals

The particle size of biosolids significantly affected the growth of corn seedlings (Figure 5-27). Five weeks following planting, the biomasses of corn plants grown with the Los Angeles biosolids-treated media were inversely correlated with particle size of the biosolids and biomass yields were significantly different at $p < 0.05$. From the < 0.83 mm fraction to the 6.35–9.91 mm fraction, the biomass yields of the Los Angeles biosolids treated plants dropped from >100 g per pot to approximately 40 g per pot. For the MWRDGC 1975 biosolids treatment, a similar trend was observed. As the N content of MWRDGC 1975 biosolids (0.8%) was considerably lower than that of the Los Angeles biosolids (4.4%), the biomass yield of the MWRDGC 1975 treatment was across the board four to five times lower than the corresponding treatment of the Los Angeles biosolids. Although the half-strength Hoagland solution used for irrigation contained adequate amounts of plant nutrients for normal growth, results of this experiment illustrated that plants preferentially absorbed nutrients from biosolids. For a given biosolids, the biomass yields were inversely related to the size of the biosolids particles in the media. This result suggests that the interaction at the root–biosolids particle interface played a significant role in the root absorption of nutrients.

The concentrations of metals in the plant tissues were affected by the concentration of metals in the biosolids and by the particle sizes of the biosolids. Metal concentrations in the Los Angeles biosolids were considerably lower than those in the MWRDGC 1975 biosolids (Table 2-2). The resulting metal contents of the plant tissues also were lower than those in the corresponding fractions of the MWRDGC 1975 biosolids treatments (Figure 5-28). In general, plants readily absorbed the Zn, Cu, and Cd from the biosolids particles, whereas the amounts of Ni, Cr, and Pb absorbed by plants were relatively small. For Zn and Cd, the total metal uptake by corn shoots inversely correlated with particle sizes of biosolids. Previous reports on the affinity of metal binding by biosolids indicated that the biosolids affinity for metals, in descending order, was Pb > Cu > Zn > Cd (Elliott et al., 1986; Harter, 1983). Chaney (1990) and Logan and Chaney (1983) concluded, after extensive comparison of data from divergent sources, that Cd and Zn had lower adsorption affinities and greater phytoavailability than Pb and Cu in biosolids-treated soils. It appeared that even for the tightly bonded elements such as Cr and Pb, plant absorption might be enhanced by the increase of specific surface area of the biosolids particles.

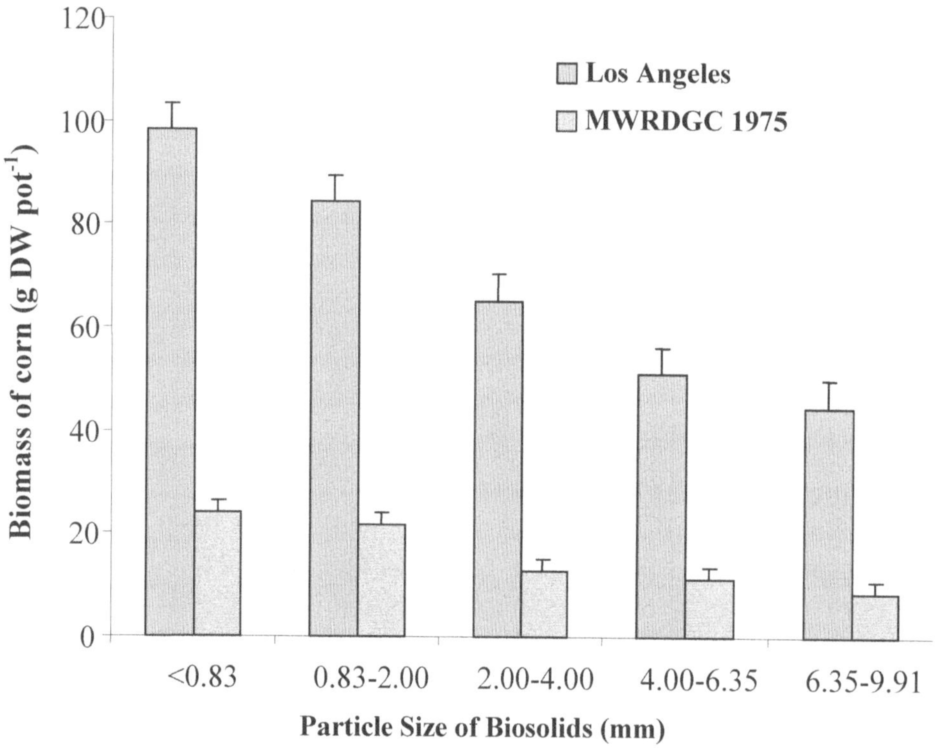

Figure 5-27. Biomass of Corn Grown on Particle Size-Classified Los Angeles and MWRDGC 1975 Biosolids.

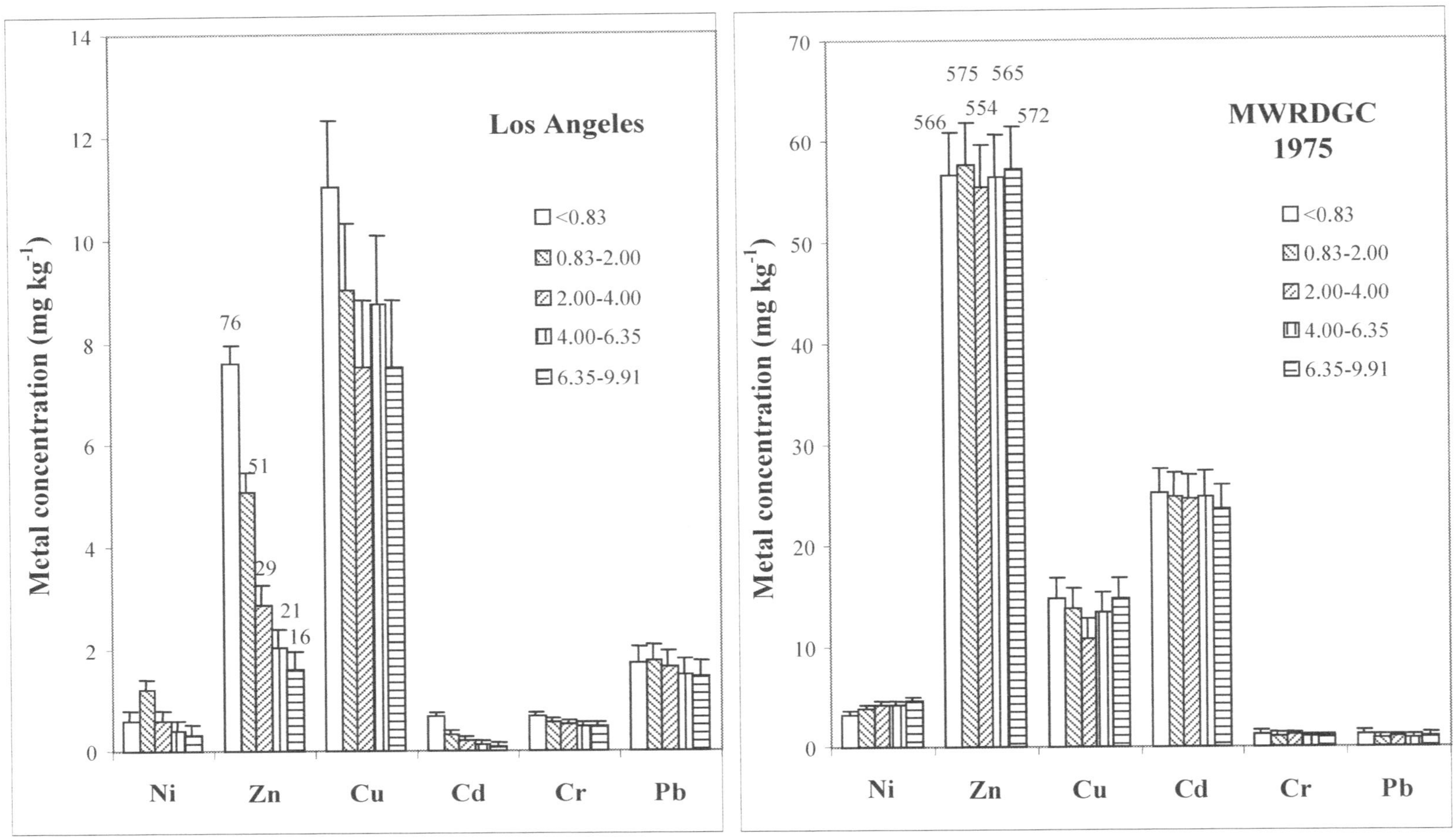

Figure 5-28. Metal Concentrations of Harvested Plant Tissue of Corn Grown on Particle Size-Classified Los Angeles and MWRDGC 1975 Biosolids. (Note: Zn was plotted on a different Scale and its values are shown directly above the bars.)

The relationship between plant metals concentrations and particle size-class differed for the two biosolids. The particle morphology and the resulting specific surface area differentiations regulated the amounts of metals absorbed by plants. For example, Zn and Cd concentrations of the harvested plant tissues treated with Los Angeles biosolids decreased as the size of the biosolids particles increased. As the total amounts of metals in the growing media were the same for all treatment levels, the reduction in metal concentration appeared to indicate that the metal uptake (metal absorbed per unit biomass) steadily decreased as the particle size of the biosolids increased. Similar conclusions may be drawn for the other elements. On the other hand, for corn grown on the MWRDGC 1975 biosolids-treated media, Cd and Zn concentrations of the harvested plant tissues were not inversely proportional to the particle sizes. In fact, the concentrations were not significantly different between the particle size treatments. Two factors appeared to influence the outcome: (1) significantly larger amounts of Cd and Zn were present in the rhizosphere of the MWRDGC 1975 biosolids-treated media and (2) the specific surface areas of all particle size fractions were essentially the same. Plant uptake is affected by the amounts of metals available and the surface area of the metal containing biosolids particles. Under these circumstances, the plant uptake would not be inhibited by the availability of Cd and Zn in the rhizosphere and the surface areas, where root came into contact with the biosolids, were essentially the same for all particle size fractions. When the Cd and Zn concentrations in plant tissues that received the MWRDGC 1975 biosolids of different particle sizes were not significantly different, it was indicative that metal uptake (as measured by the metal concentration of the plant tissue) by plants was inhibited by neither the availability of metals in the rhizosphere nor the surface areas of the biosolids particles that might limit the access of the available metals to plants. The biomass yields in this case responded to the inverse particle size classification. As the size-classified MWRDGC 1975 biosolids also had the same amounts of N and P, the difference in the biomass yields for size-classified MWRDGC 1975 biosolids may be attributed to the difference in their surface areas. It appears, therefore, that for biosolids high in metal contents (i.e., MWRDGC 1975), the plant uptake is determined by the abundance of the metal elements.

The total metal uptake by plants is a function of both the biomass yield and the metal concentrations in plant tissues. The total metal uptake by plants grown with both biosolids was inversely classified with the particle sizes of biosolids (Figure 5-29).

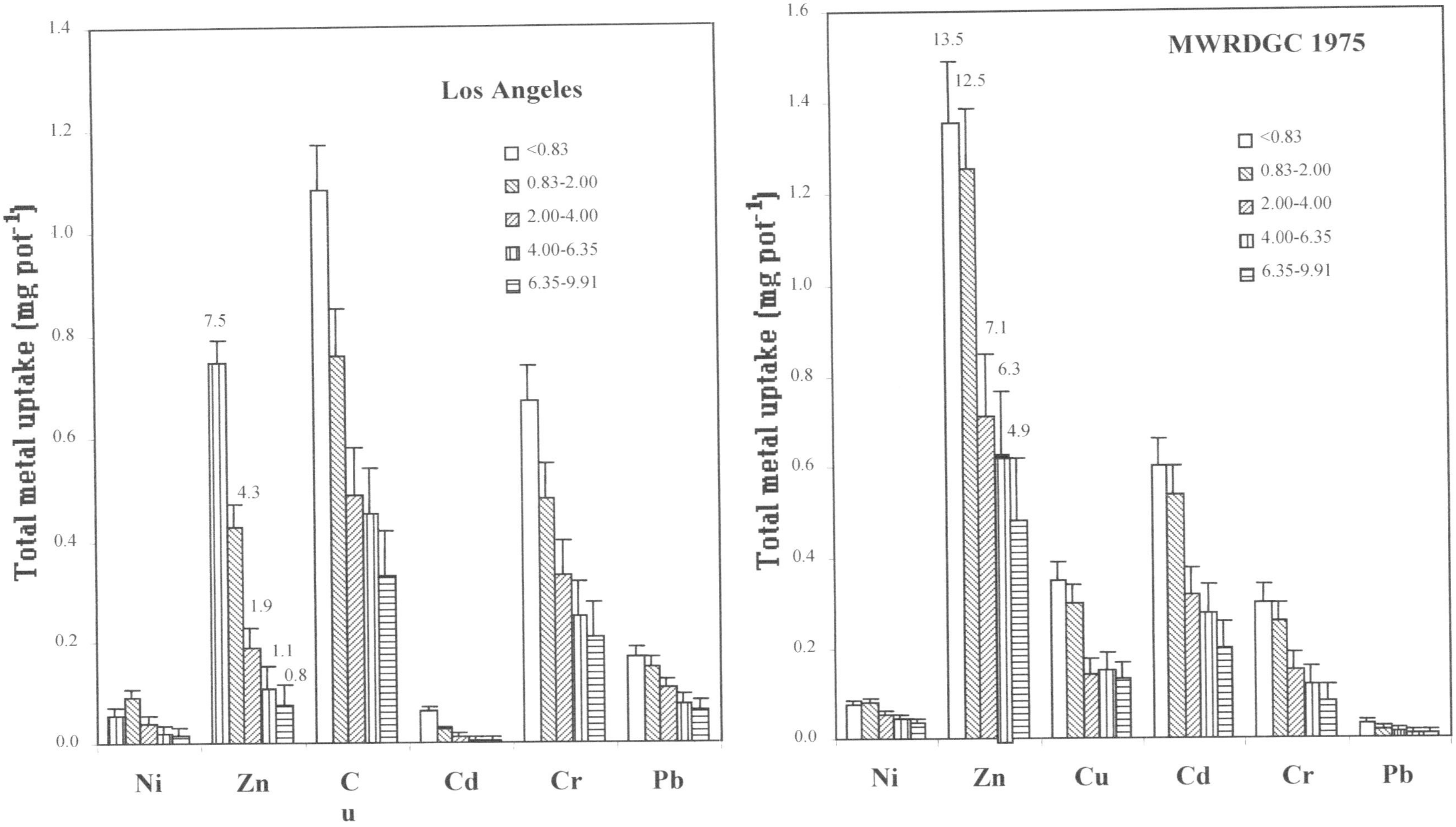

Figure 5-29. Total Plant Uptake of Metals by Corn Grown on Particle Size Classified Los Angeles and MWRDGC 1975 Biosolids. (Note: Zn was plotted on a different Scale and its values are shown directly above the bars.)

The particle size distribution of biosolids can be affected by the chemical additives used in the dewatering process and the subsequent post-dewatering process train. Frequently, the finished biosolids were mechanically homogenized to produce a consistent and uniform product. These processes will have implications on plant absorption of heavy metals when biosolids are land-applied.

5.5.3 Particle Size Effects - Conclusions

1. Using scanning electron microscopy, two distinctive types of biosolids particles were identified. One type of particles, represented by the MWRDGC 1975 biosolids, was aggregated, porous and spongy, with rough exterior surfaces and interconnected internal passages. The specific surface areas of these particles did not change significantly with particle size. The second type of particles, represented by the Los Angeles biosolids, was discrete, granular, and blocky, with flat surfaces and few exterior features. For comparable size ranges, the specific surface areas of this second type of particles were significantly smaller that the particles of the first type.
2. Particle size classification, however, was not correlated with the nitrogen and metal contents of the biosolids; for biosolids from the same source, the concentrations were essentially the same in all particle size groups.
3. Biomass yields of corn grown on the size-classified biosolids-treated media were inversely related to the particle size for both types of biosolids.
4. For the Los Angeles biosolids (discrete, granular, and blocky), the metal concentrations in the plant tissue produced on the size-classified biosolids treatments were inversely related to the particle sizes. Smaller particles with greater specific surface area were associated with higher rates of metal uptake.
5. For the MWRDGC 1975 biosolids (aggregated, porous and spongy), the metal concentrations in the plant tissues were not significantly different due to the particle size classification. The abundance of available metals in the biosolids and/or abundance of sites for interfacial reactions (i.e., surface area) might be the causes. The N contents of the size-classified biosolids were not different, yet the biomass yields of corn were inversely related to biosolids particle size. Thus, it appears that the particle size classification was not a factor causing the high metal concentrations in plant tissue and, instead, the metal concentrations in the biosolids were the primarily contributing factor.
6. When the biomass yields and the plant tissue concentrations were considered collectively, the particle size of the biosolids affects the plant's ability to absorb biosolids-borne heavy metals.

CHAPTER 6.0

ROOT EXUDATES

Plant roots secrete a variety of soluble substances that include sugars, amino acids, organic acids, lipids, coumarins, flavonoids, proteins, enzymes, aliphatics, aromatics, and low molecular weight carbohydrates (Rovira, 1969; Fan et al., 1997; Schilling et al., 1998; Pomilio et al., 2000). Among these substances, organic acids are considered to be the most abundant and chemically active root exudate component in the rhizosphere. The organic acids were responsible for the metal dissolution in the rhizosphere and subsequent uptake by plants. Quantitative measurements indicated that organic acid concentrations as high as 50 mM were found within 1 mm of the root surface (Smith, 1970; Barber and Gunn, 1974; Jones et al., 1996b), with typical concentrations in the range of 10-20 mM. Other reports showed that organic solutes present in root cells existed primarily as amino acids (1020 mM) and sugars (90 mM) (Jones and Darrah, 1994,1996).

In the rhizosphere, root exudates are susceptible to modifications to secondary products—namely organic acids—by soil microorganisms (Anderson et al., 1993). Microbial metabolism in the rhizosphere may utilize the carboxylic acids as substrates or produce more organic acids as byproducts. Kraffczyk et al. (1984) reported that root exudates accumulating under aseptic conditions contained higher concentrations of monosaccharides than those collected under septic conditions, while the concentrations of organic acids were similar under both conditions. Since microorganisms preferentially utilize the carbohydrates in root exudates (Jones and Darrah, 1994; Jones et al., 1996a), organic acids are likely to be the prevalent chemical substances in root exudates (Inskeep and Comfort, 1986).

Root exudates were found to increase the availability of nutrients (Romheld and Marschner, 1986; Fan et al., 1997) and enhance the trace element uptake by plants (Welch, 1995). Among the exuded components, amino acids and sugars appeared to have little or no role in the complexing or mobilizing of plant nutrients (Jones et al., 1994; Jones and Darrah, 1996). Organic acids are the most abundant and chemically active component of exudates in the rhizosphere (Jones and Darrah, 1994; Jones et al., 1996a, 1996b) and they may mobilize mineral nutrients by acidifying the rhizospheric soils, thus solubilizing adsorbed and/or precipitated elements. Organic acids and metals also may form soluble complexes, thus increasing the plant availability of metals (Jauregui and Reisenauer, 1982; Mench and Martin, 1991). Interactions of root exudates with Fe, Mn, Cu, and Zn have been examined (Treeby et al., 1989; Zhang et al., 1991), and effects of organic acids on P dissolution in soils also have been reported (Dinkelaker et al., 1989). The commonly found organic acids in the rhizosphere included acetic, lactic, malic, malonic, acetic, citric, fumaric, succinic, and tartaric (Laheurte et al., 1990; Otani et al., 1996; Young et al., 1998). Studies on mineralization of root exudates indicated that citric and maleic acids in the rhizosphere are rapidly degraded or converted to other organic acids (Jones and Darrah, 1994). Oxalic and malic acids are present primarily in cells for cation charge balancing or for maintaining osmotic potential and are much harder to degrade. Otani et al. (1996) and Ahumada et al. (1999, 2001) reported that lactic, acetic, and butyric acids were the most common organic acids in the rhizosphere of forest and contaminated soils.

Environmental factors may affect the types and amounts of organic acids in soils. They include the nutritional stress experienced by plants, such as deficiencies of P, Zn, and Fe, (Kato-Noguchi et al., 1994; Yang and Crowley, 2000; Rengel and Romheld, 2000), the presence of microorganisms in the growth medium (Uren and Reisenauer, 1988; Meharg and Killham, 1991; Kozdroj and Van Elsas, 2000), soil temperature (Hale et al., 1978; Uselman et al., 2000; Pramanik et al., 2000), the imbalance in plant absorption of cation and anion (Chang and Roberts, 1991; Marschner, 1995; Ali et al., 2000), soil texture (Burken and Schnoor, 1996; Iijima et al., 2000), soil moisture (Nye et al., 1977), plant species and cultivars (Szmigielska et al., 1995; Bozhkov et al., 1996), age of root (Schilling et al., 1998), salinity (Perez-Alfocea et al., 2000), and light intensity (Kato-Noguchi et al., 1994; Pramanik et al., 2000).

The soil is the entry point for potentially harmful substances to enter the human food chain. When municipal biosolids are applied on cropland, potentially harmful trace metal elements are inadvertently introduced into the root zone along with the organic carbon and essential nutrients. The presence of biosolids in the soil is expected to influence the production of the organic acids in the rhizosphere of plants grown on biosolids-treated media. In turn, the plant absorption of toxic metal elements also will be affected. The purpose of the studies summarized in this chapter is to evaluate effects of biosolids on the organic acid production in the rhizosphere.

6.1 Organic Acids Collected Under Aseptic and Septic Conditions

Microbial activity in the rhizosphere might affect the amounts and composition of root exudates. We compared the volume and composition of root exudates collected under aseptic and septic conditions. At the time of root exudate collection, the pots were disconnected and removed from the hydroponic systems, the rhizosphere was flushed to remove residual nutrients, and plants were allowed to photosynthesize for five hours. For aseptic collection, the antimicrobial agent, Micropur, was added at the final flushing to eliminate the microbial activity in the rhizosphere before the commencement of the photosynthetic period.

The microbial activity in the rhizosphere did not significantly affect the amounts of organic acids produced when plants were grown on the same media (Figure 6-1). Except for small amounts of glutaric and succinic acids, plants produced the same organic acids in both aseptic or septic collection conditions in both standard nutrient solution media and biosolids-treated media (Table 6-1). For the standard plant nutrient media, lactic, acetic, butyric, and tartaric acids accounted for 91 and 89% of the total organic acids collected under aseptic and septic conditions, respectively. For the biosolids-treated media, the same four organic acids accounted for 92 and 95% of the total organic acids collected under aseptic and septic conditions (Table 6-1). When the root exudates of plants grown on standard plant nutrient media were collected under septic conditions, butyric acid in the root exudates proportionally increased and the lactic, acetic, and tartaric acids decreased in comparison with that of the root exudates collected aseptically. In the biosolids-treated media, the acetic and butyric acids in the root exudates collected under septic conditions increased while the lactic and tartaric acids decreased.

Table 6-1. Mole-based Fractional Composition of Two-week Old Corn Root Exudates Collected Under Septic and Aseptic Conditions.

Treatment	Growth Media	Organic Acid (Fraction of Total COO^-)						
		Lactic	Acetic	Butyric	Glutaric	Succinic	Tartaric	Oxalic
Septic	Standard Nutrient	0.01	0.11	0.53	0.07	n.d.[a]	0.23	0.05
	Biosolids-Treated	0.30	0.30	0.20	n.d.	0.05	0.15	0.01
Aseptic	Standard Nutrient	0.13	0.32	0.08	0.02	n.d.	0.38	0.06
	Biosolids-Treated	0.44	0.13	0.04	n.d.	0.04	0.35	0.01

[a]n.d., not detected

6.2 Organic Acids Production

Organic acids were recovered in the media of all treatments two weeks into the experiment, including the treatment blanks where no plant was cultivated (Figure 6-2). More organic acids were recovered from media where biosolids were added than the corresponding treatments that received only the standard nutrient solution. The presence of plant roots further enhanced the organic acid production. In all, the biosolids-treated and planted media produced the most organic acids. This trend persisted for the entire growing period and, for a given treatment, the amount of organic acids recovered did not change significantly ($p < 0.01$) over the first 12 weeks of the growing period. The average amounts of organic acids recovered (± SD) were 14.03 ± 2.36, 6.11 ± 1.63, 3.56 ± 0.94, and 2.69 ± 0.69 x 10^{-3} moles of COO^- per pot per collection period for planted biosolids, planted standard nutrient solution-treated, blank biosolids, and blank standard nutrient solution treatments, respectively (Figure 6-3). The collection period referred to is five hours of photosynthesis following flushing the growth media with deionized water.

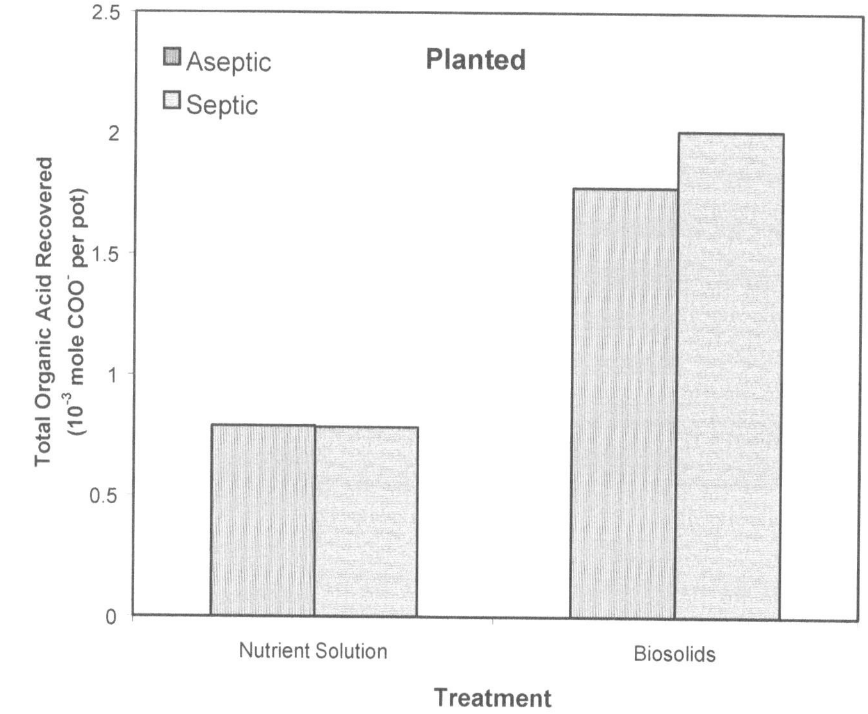

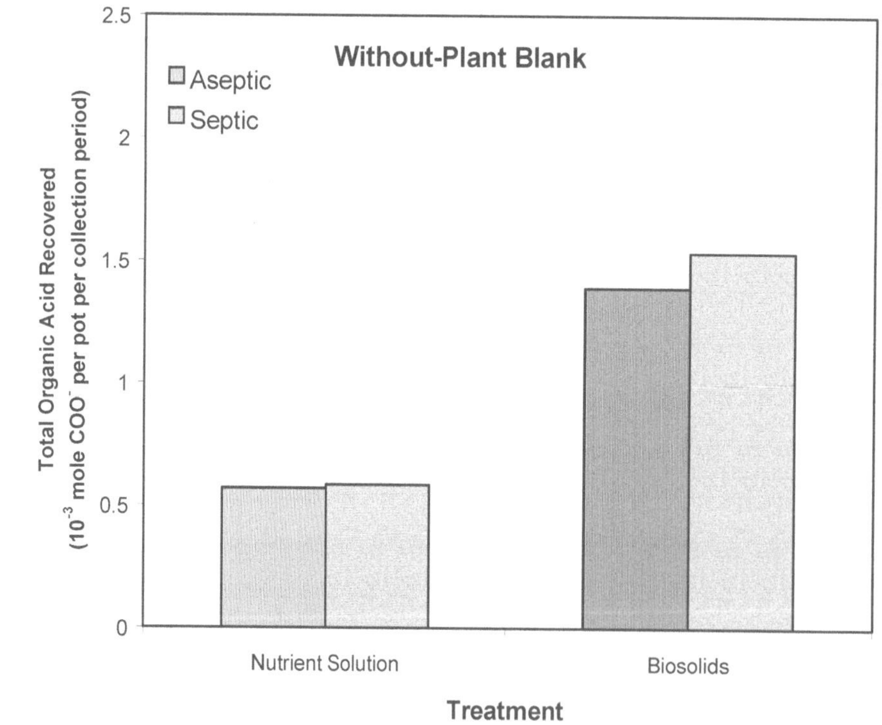

Figure 6-1. Total Organic Acid Recovery Under Septic and Aseptic Conditions. (Collection period refers to 5 hours of photosynthesis following flushing growth media with deionized water.)

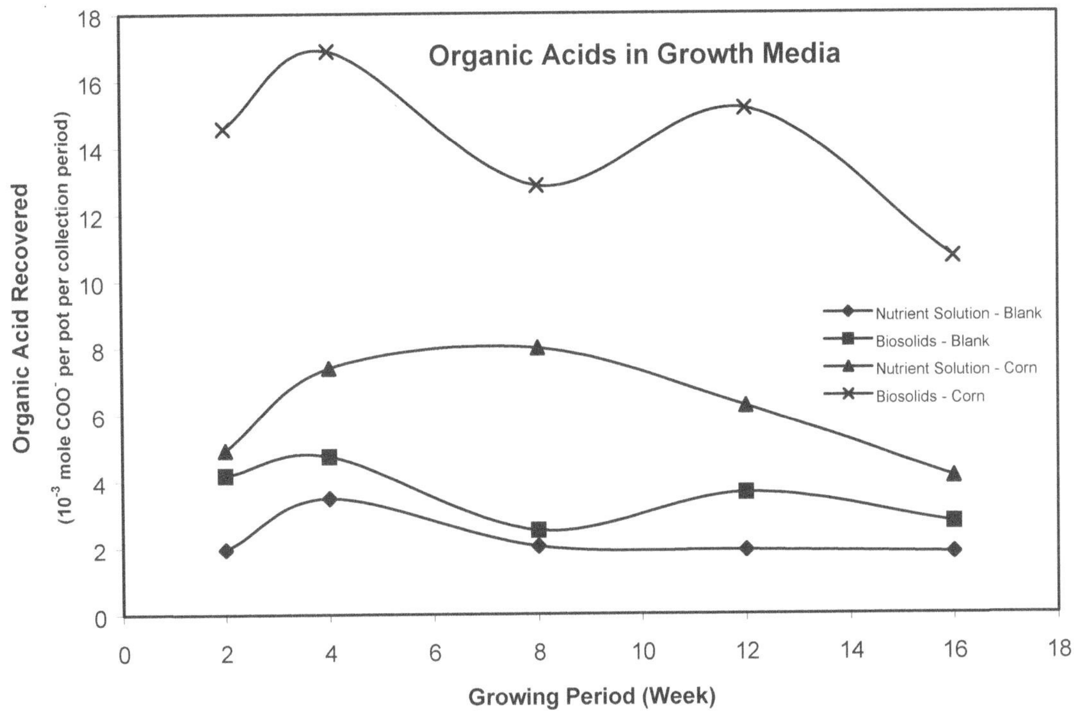

Figure 6-2. Amounts of Organic Acids Recovered from Plant Cultivating Media and Corresponding Blank Treatments. (Collection period refers to 5 hours of photosynthesis following flushing growth media with deionized water.)

Under septic conditions, spontaneous microbial activities would contribute to the organic acid production. The recovery of organic acids in the blank treatment was an indication that much of the organic acids were byproducts of microbial metabolism. In the nutrient solution blank treatment, there would not be organic substrate to induce the microbial metabolism. The growth and subsequent degradation of algae and chemotrophic bacteria likely initiated the process. When the plants were grown, root exudates provided the substrates that enhanced microbial metabolism, and thus organic acid production. Biosolids were the additional substrate that induced even higher levels of microbial activities. Therefore, the largest amounts of organic acid production were in biosolids-treated media where plants were grown.

Smith (1970) and Schilling et al. (1998) showed that organic acids were secreted more abundantly during the initial stages of plant growth, with primary roots exuding more than mycorrhizal roots. The fact that the volume of organic acids recovered did not increase significantly over the period of growth was another indication that significant amounts of the organic acids in the rhizosphere were the result of microbial activity and their production was in proportion to the availability of substrates and was subject to the limitation of the media volume.

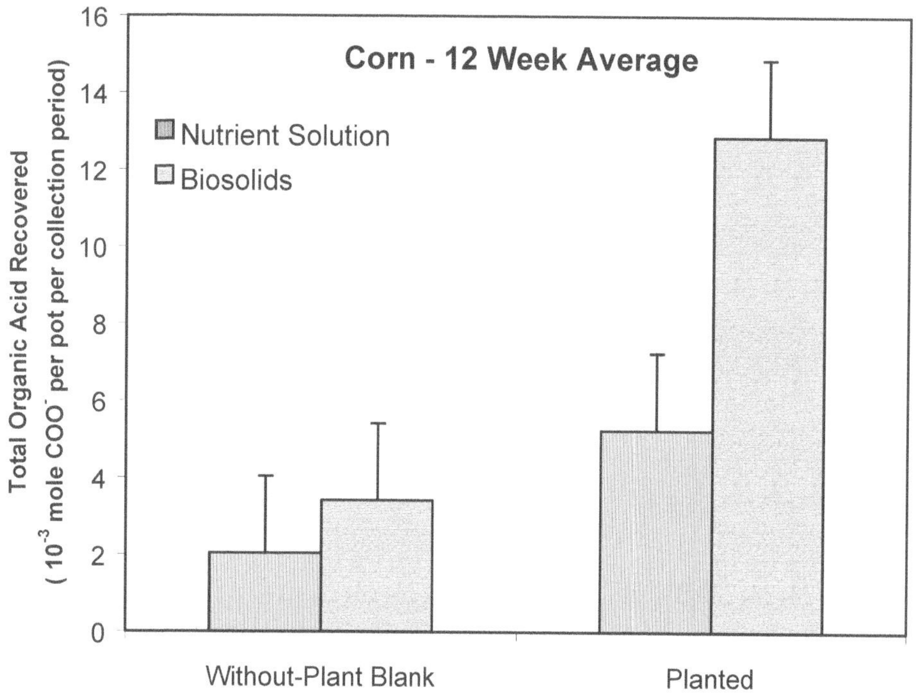

Figure 6-3. Organic Acid Production of Corn Grown from Seedling to Maturity (Average ± SD).

6.3 Organic Acid Composition

Four organic acids—lactic, acetic, butyric and oxalic acids—were found in the blank treatments, i.e., sand media and biosolids-treated sand media that were irrigated with nutrient solutions, but not planted (Figures 6-4 and 6-5). While the biosolids-treated blank media produced more organic acids than the nutrient solution only blank (Figure 6-2), the relative proportions of each acid in the mixtures were similar in these two treatments. Generally, lactic, acetic, butyric, and oxalic acids accounted for 0.4-0.5, 0.2-0.3, 0.2-0.3, and 0.05-0.1 mole fractions of the COO^-, respectively. During the growing season, the COO^- mole fraction of each acid fluctuated dynamically, possibly in response to the change in substrate availability and microbial population dynamics (Figures 6-4 and 6-5). In Figures 6-4 and 6-5, and subsequent figures, the organic acids recovery over the growing seasons is plotted on the same scale to illustrate the magnitude of changes due to the treatments. The average organic acid composition of the blank media is shown in Figure 6-6.

When the plants were grown, 9 to 11 organic acids were found in the solution recovered from the rhizosphere. In addition to the four organic acids present in the blank media, tartaric, maleic, succinic, valeric, glutaric, pyruvic, and propionic acids were detected (Figures 6-7 and 6-8). By associations, the additional organic acids, which were higher in molecular weights and some of which contain two carboxyl functional groups, were related to excretions from the roots. Among the organic acids found, lactic, acetic, and butyric acids were always the largest in quantities. Individually, they might fluctuate over time from 0.15-0.35 of the COO^- mole fraction. Collectively, they accounted for 0.7-0.85 of the COO^- mole fraction of the organic acids recovered at any given time. Next in abundance were oxalic, maleic, and tartaric acids, whose COO^- mole fractions varied from 0.05-0.1. There were even smaller amounts of succinic, valeric, glutaric, propionic, and pyruvic acids. Their presence in the organic acid mixture usually was <0.05 in COO^- mole fraction each. Succinic, glutaric, and valeric acids were always present.

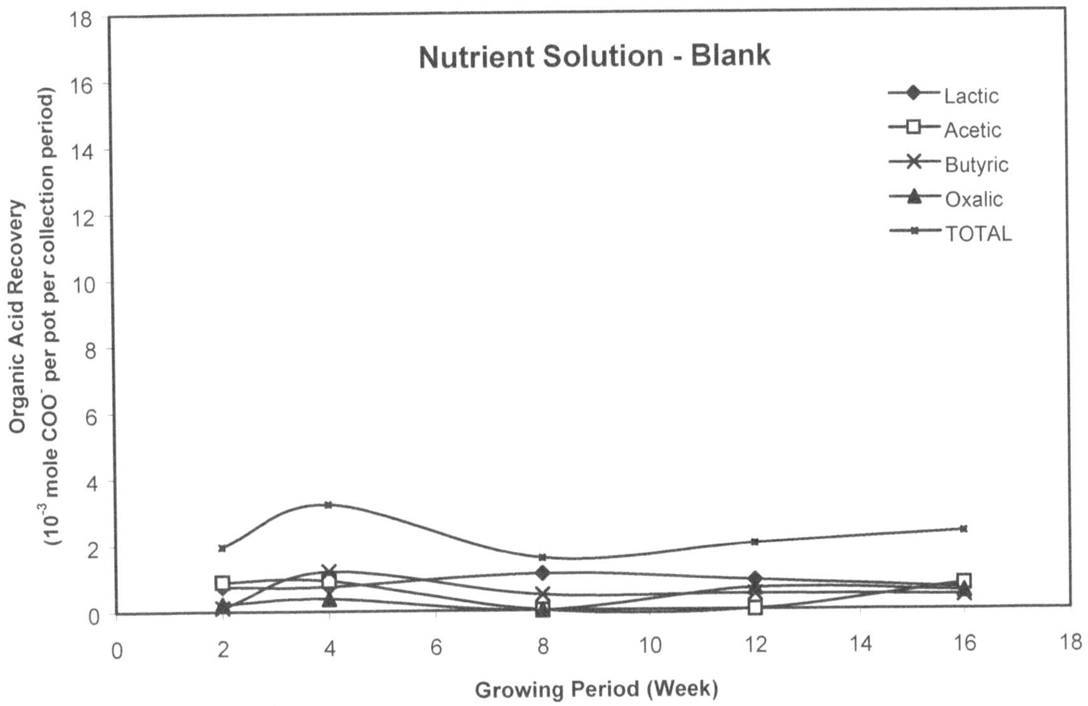

Figure 6-4. Dynamic Change in Organic Acid Composition in the Nutrient Solution Irrigated Blank Media.

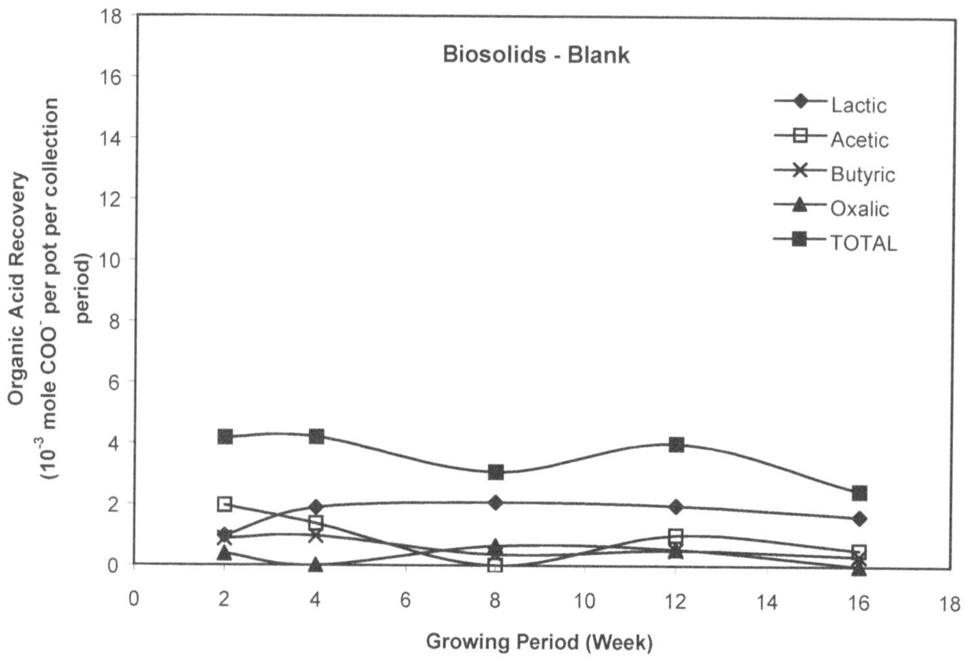

Figure 6-5. Dynamic Changes in Organic Acid Composition in the Biosolids-Treated Blank Media.

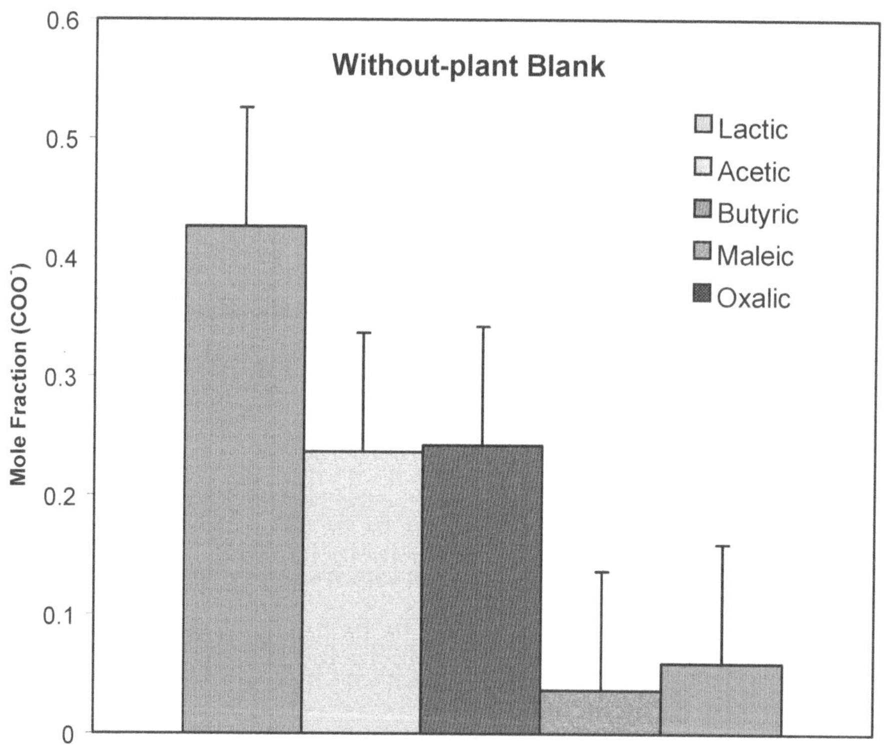

Figure 6-6. Twelve-Week Average Organic Acids Composition of Nutrient Solution-Irrigated Blank Media.

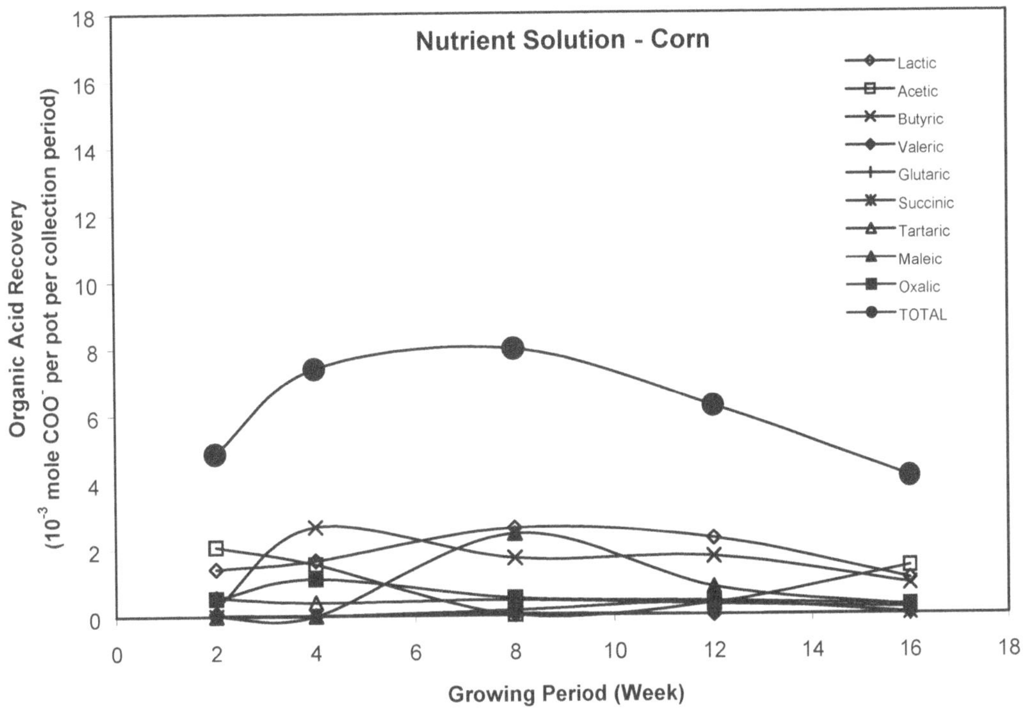

Figure 6-7. Dynamic Changes in Organic Acid Composition in the Nutrient Solution-Treated Rhizosphere of Corn.

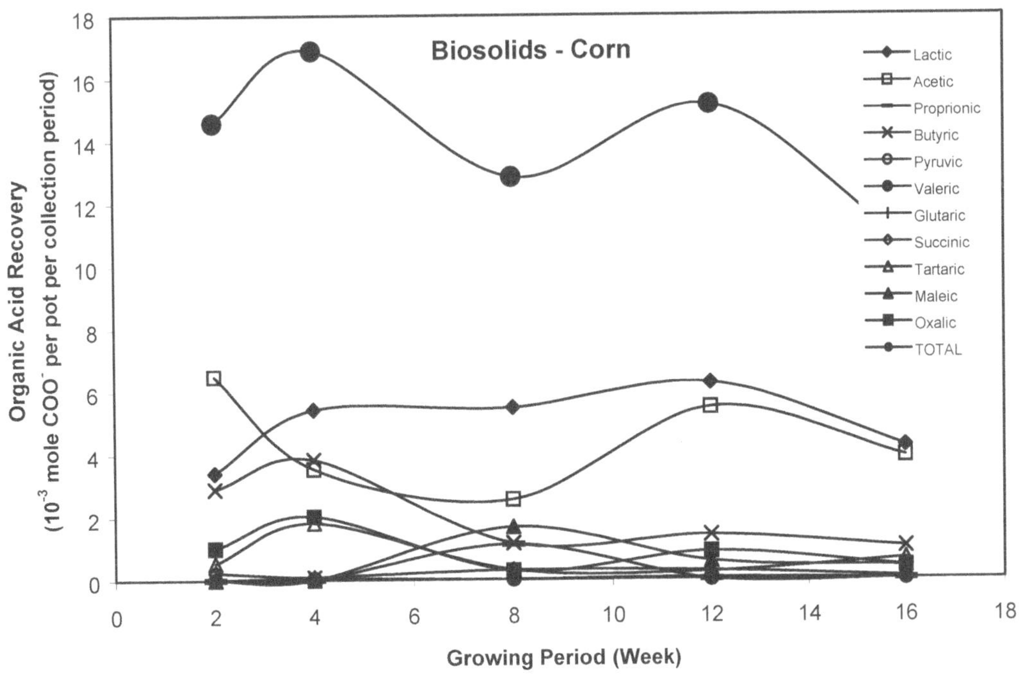

Figure 6-8. Dynamic Changes in Organic Acid Composition in the Biosolids-Treated Rhizosphere of Corn.

Propionic and pyruvic acids were found some of the times and only in the rhizosphere of the media treated with biosolids. The effect of biosolids on organic acid compositions of the rhizosphere appeared slight. The quantity of individual acids in the rhizosphere dynamically fluctuated during the growing season. The relative composition of these three groups of acids remained the same throughout the growing season. The concentration of citric acid, an essential component of plants and a commonly reported organic acid of root exudates, was below the detection limit in essentially all of the determinations. Studies showed that, under septic conditions, citric and maleic acids added into the rhizosphere or bulk soils at concentrations typically found in the root exudates, were rapidly degraded or converted to other organic acids of the Krebs cycle (Jones and Darrah, 1994; Jones et al., 1996a) by microbial metabolism.

When the organic acid compositions of the planted (Figure 6-9) and the blank media (Figure 6-6) were compared, it was apparent that lactic, acetic, butyric—and to a certain extent oxalic—acids in the rhizosphere were associated primarily with microbial metabolism and their quantities were far greater than glutaric, maleic, propionic, pyruvic, succinic, tartaric, and valeric acids in the rhizosphere that were induced by the growth of plants, most likely from excretion by the root (Figure 6-9).

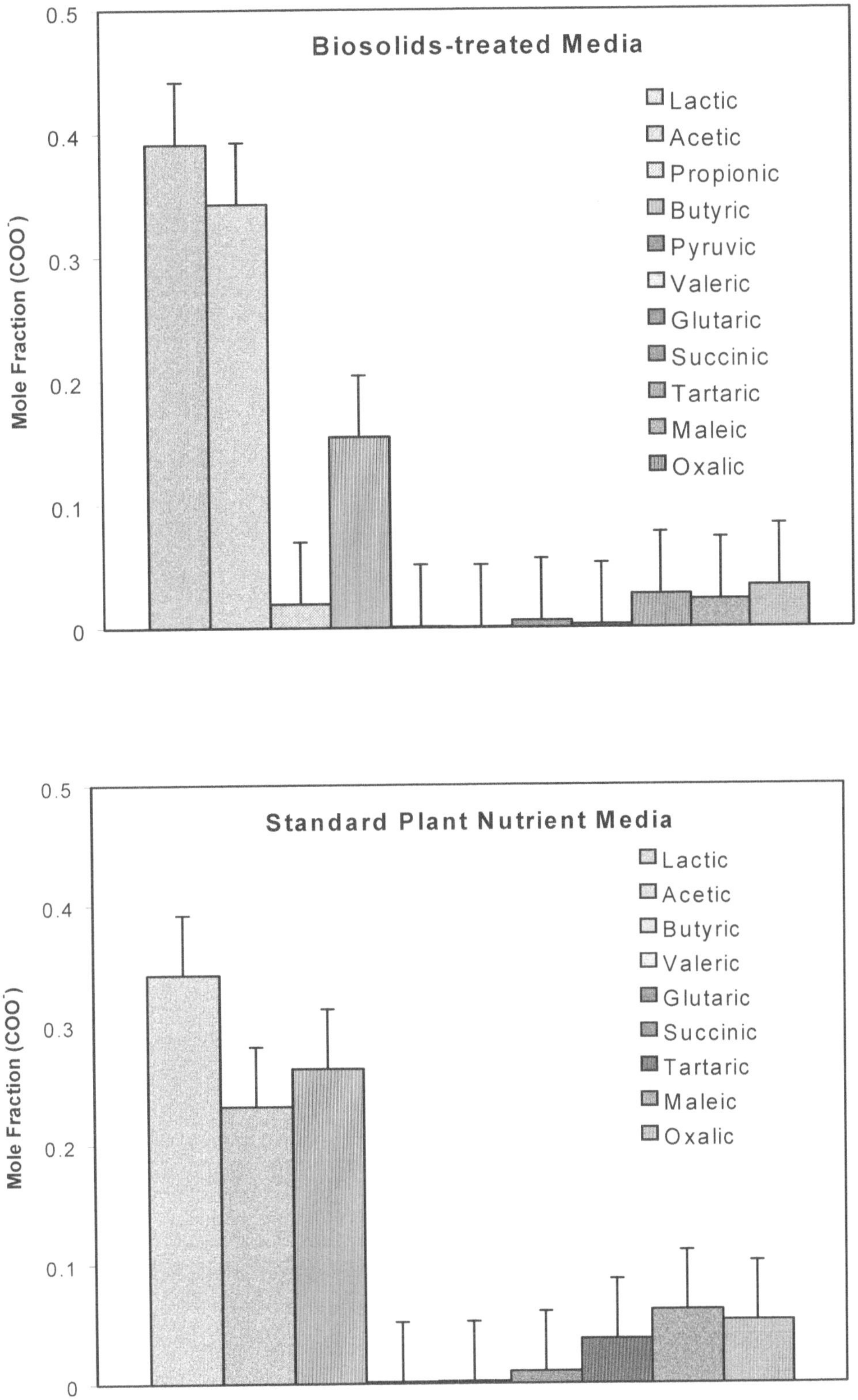

Figure 6-9. Organic Acids Composition of Root Exudates of Corn Grown from Seedling to Maturity (Average ± SD).

The aggregated organic acid composition of root exudates of corn is summarized in Table 6-2.

Table 6-2. Organic Acids in Root Exudates of Corn Grown on Standard and Biosolids-Treated Media.

Time	Standard Plant Nutrient Medium			Biosolids-Treated Medium		
	Primary[a]	Secondary[b]	Trace[c]	Primary	Secondary	Trace
Week 2	Lactic Acetic Butyric	Glutaric Tartaric Oxalic	—	Lactic Acetic Butyric	Succinic Tartaric Oxalic	—
Week 4	Lactic Acetic Butyric	Tartaric Oxalic	Glutaric	Lactic Acetic Butyric	Glutaric Tartaric Oxalic	Valeric Succinic
Week 8	Lactic Butyric Maleic	Succinic Tartaric Oxalic	Acetic Valeric Glutaric	Lactic Acetic Butyric	Propionic Glutaric Tartaric Maleic Oxalic	—
Week 12	Lactic Butyric	Acetic Succinic Tartaric Maleic Oxalic	—	Lactic Acetic Butyric	Pyruvic Glutaric Succinic Tartaric Maleic Oxalic	—
Week 16	Lactic Acetic Butyric	Tartaric Maleic Oxalic	Valeric Succinic	Lactic Acetic Butyric	Tartaric Maleic Oxalic	Glutaric

[a]Organic acids that accounts for > 10% of the total detected organic acids (COO^-) in the rhizosphere.
[b]Organic acids that accounts for 1 to 10% of the total detected organic acids (COO^-) in the rhizosphere.
[c]Organic acids that accounts for < 1% of the total detected organic acids (COO^-) in the rhizosphere.
– Not detected.

Based on the organic acid compositions, there are 1.26 and 1.46 moles of COO^- ligand to complex metals in each COO^- mole of mixed organic acids in the root exudates of corn grown in standard and biosolids-treated media, respectively. As the biosolids treatment enhanced organic acids production, the COO^- ligand present in the biosolids-treated media would be three to four times greater than those of the standard nutrient media. The lactic, butyric, tartaric, maleic, and oxalic acids account for 84% of metal-complexing capacity of root exudates in standard nutrient solution sand media. The lactic, acetic, butyric, tartaric, and maleic acids account for 87% of the metal-complexing capacity of the root exudates of corn grown on biosolids-treated media. These data are used as a basis for developing the synthetic root exudates to study root exudate-specific metal solubility. At this time, concentrations of organic acids in the rhizosphere cannot be accurately confirmed. However, they may be approximated by normalizing the organic acids collected to the water content of the sand media and to the estimated volume of root exudates produced. Based on an estimate of 200 mL of water retained per pot in the sand media, the organic acid concentrations would be approximately 2 and 20 mM in the solutions in the standard and biosolids-treated media, respectively. Concentrations may be further elevated in the rhizosphere solution by a factor of 10.

6.4 Root Exudates of Six Plant Species

To compare the root exudates of different plant species, six species of plants were grown on nutrient solution irrigated sand media, biosolids-treated sand media, and rock phosphorus-treated sand media. For plants of the same species, the biosolids-treated media produced the largest amount of organic acids, followed in descending order by the rock phosphorus and the standard plant nutrient treatments (Figure 6-10). The organic acids in the root exudates of canola, chickpea, Swiss chard, corn, Durum wheat, and Sudan grass also were dominated by the microbial metabolism-induced organic acids in the media (Figures 6-11 to 6-16). When biosolids and phosphorus rock were present in the media, acetic acid became the most prominent component of the organic acids in the root exudates.

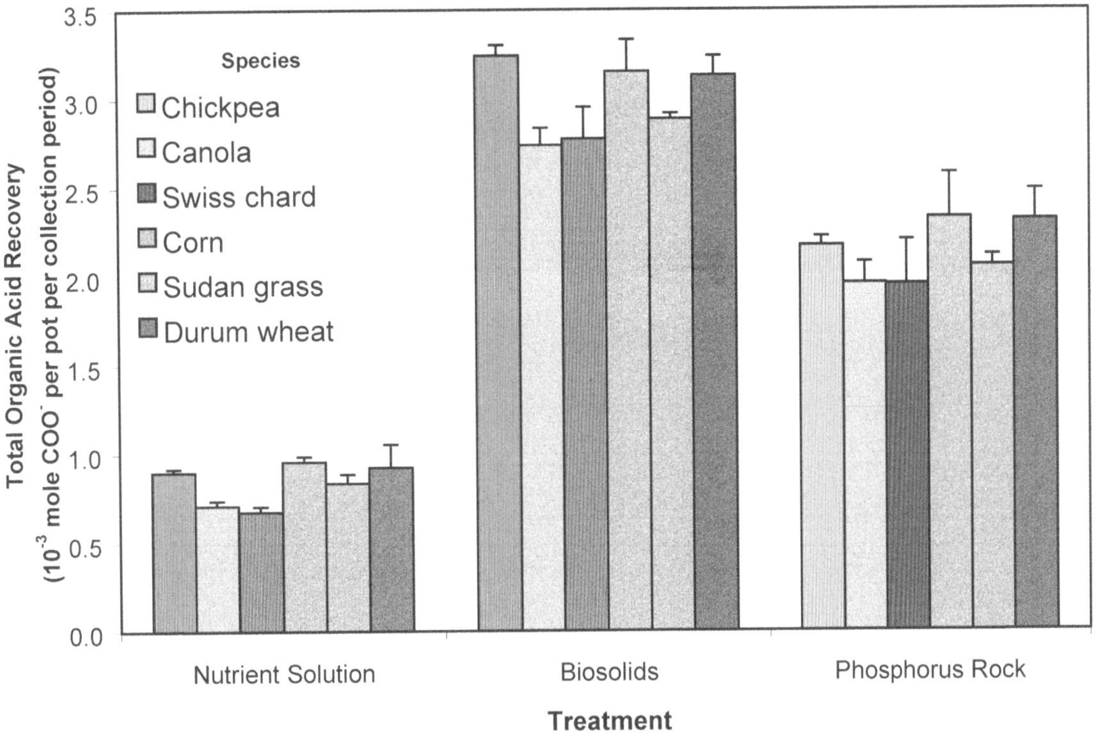

Figure 6-10. Amounts of Organic Acids Recovered from the Rhizosphere of Six Plant Species Grown on Standard Nutrient Solution-, Biosolids-, and Phosphorus Rock-Treated Media (Average ± SD).

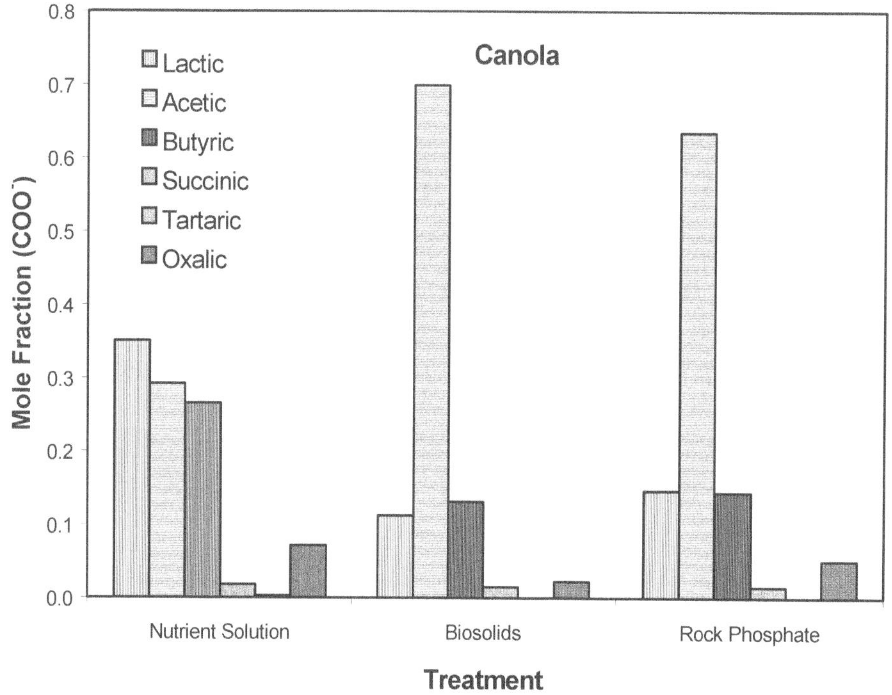

Figure 6-11. Organic Acid Composition in the Rhizosphere of Canola Grown on Three Different Media for Four Weeks.

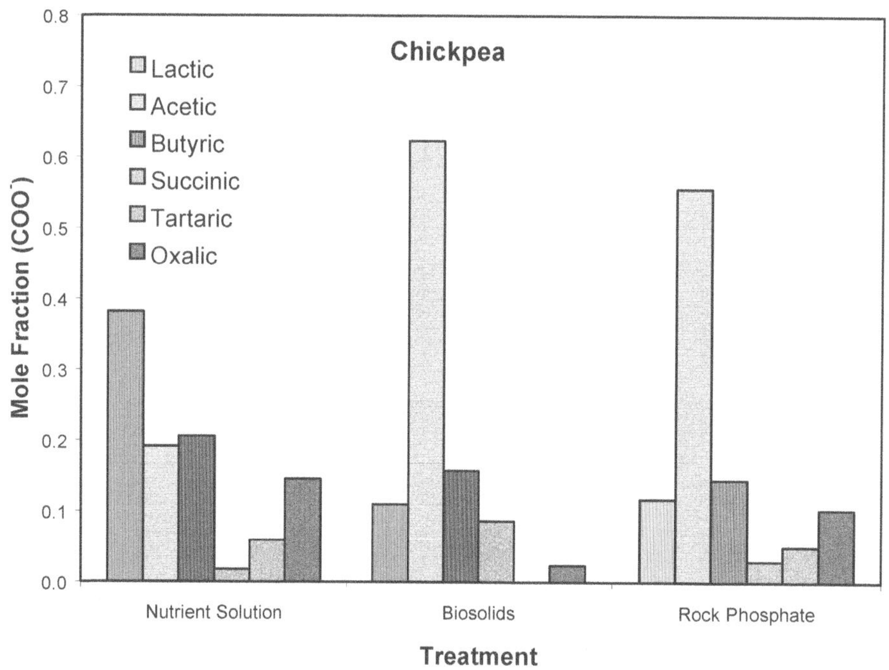

Figure 6-12. Organic Acid Composition of the Rhizosphere of Chickpea Grown on Three Different Media for Four Weeks.

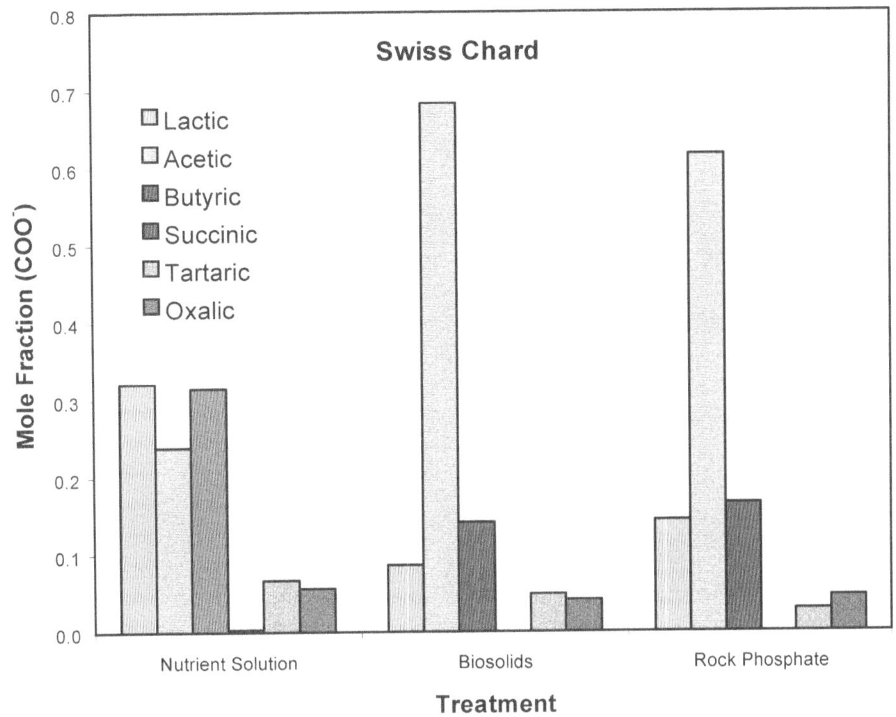

Figure 6-13. Organic Acid Composition of the Rhizosphere of Swiss Chard Grown on Three Different Media for Four Weeks.

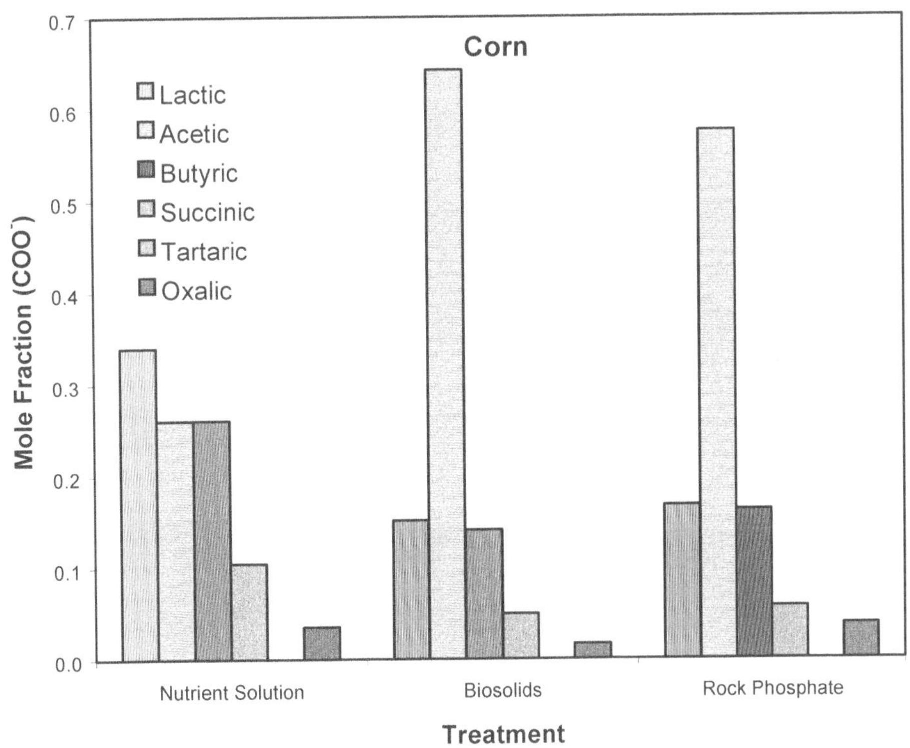

Figure 6-14. Organic Acid Composition of the Rhizosphere of Corn Grown on Three Different Media for Four Weeks.

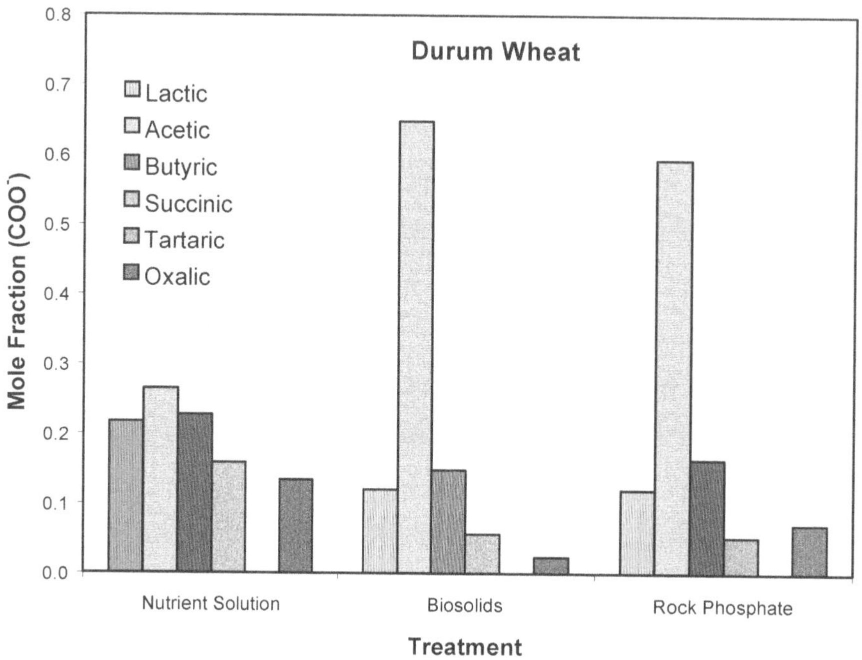

Figure 6-15. Organic Acid Composition of the Rhizosphere of Durum Wheat Grown on Three Different Media for Four Weeks.

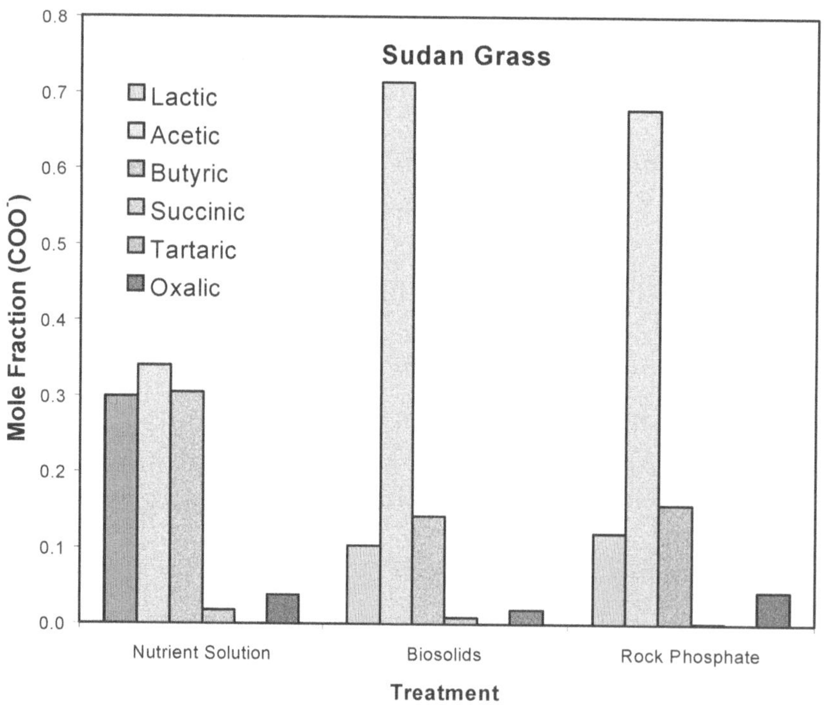

Figure 6-16. Organic Acid Composition of the Rhizosphere of Sudan Grass Grown on Three Different Media for Four Weeks.

The organic acid compositions of the plant species appeared to be a function of the growing media and the organic acid distribution patterns changed with the amendments in the media, namely biosolids or phosphorus rock. The organic acids in the root exudates again may be categorized into three groups, namely primary (>10% of the total), secondary (1-10% to the total), and trace (<1% of the total) (Table 6-3).

Table 6-3. Organic Acids of Chickpea, Swiss Chard, Canola, Durum Wheat, Corn, and Sudan Grass Grown on Nutrient Solution and Biosolids-Treated Sand Media.

Plant Species	Standard Medium			Biosolids-Treated Medium		
	Primary[a]	Secondary[b]	Trace[c]	Primary	Secondary	Trace
Chickpea	Lactic Acetic Butyric Oxalic	Succinic Tartaric	—	Lactic Acetic Butyric	Succinic Oxalic	—
Swiss chard	Lactic Acetic Butyric	Tartaric Oxalic	—	Acetic Butyric	Lactic Tartaric Oxalic	—
Canola	Lactic Acetic Butyric	Succinic Oxalic	Tartaric	Lactic Acetic Butyric	Pyruvic Succinic Oxalic	Glutaric
Durum wheat	Lactic Acetic Butyric Succinic Oxalic	Tartaric Maleic	—	Lactic Acetic Butyric	Succinic Oxalic	Glutaric
Corn	Lactic Acetic Butyric	Tartaric Oxalic	Succinic	Lactic Acetic Butyric	Tartaric Oxalic	Valeric Glutaric Succinic
Sudan grass	Lactic Acetic Butyric	Succinic Oxalic	—	Lactic Acetic Butyric	Propionic Oxalic	Pyruvic Succinic

[a]Organic acids that accounts for > 10% of the total detected organic acids (COO^-) in the rhizosphere.
[b]Organic acids that accounts for 1 – 10% of the total detected organic acids (COO^-) in the rhizosphere.
[c]Organic acids that accounts for < 1% of the total detected organic acids (COO^-) in the rhizosphere.
– Not detected.

6.5 Organic Acids of Root Exudates for Three Wheat Varieties

Three varieties of wheat—Chinese spring, Yecora Rojo and Durum wheat—were grown to determine effects of the plant variety on production of organic acids. Root exudates were extracted at the end of Weeks 4, 8, and 12 following planting.

Under the same treatment, the root exudate production might vary considerably. The collection of root exudates from Durum wheat was replicated four times and the average and the standard deviation of the root exudate production were determined (Figure 6-17). Results illustrated the extent of the variation in the amounts of organic acids recovered over the 12-week growing period. The coefficients of variation varied from 3-26%.

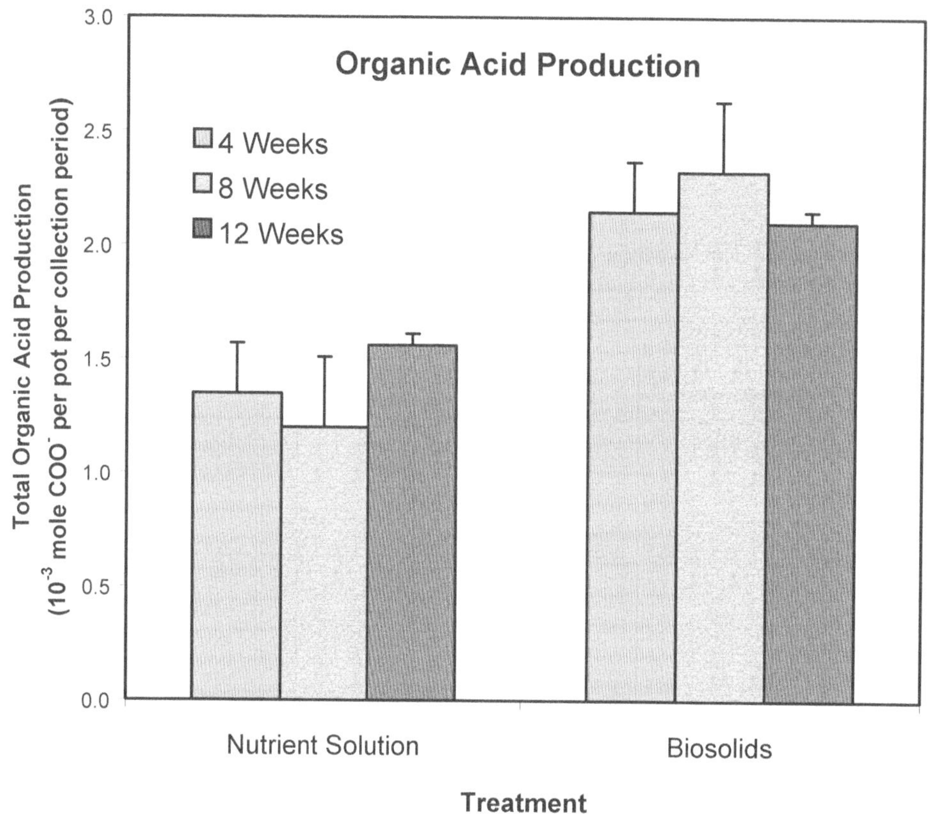

Figure 6-17. Organic Acid Produced by Durum Wheat at Different Stages of the Growing Season (Average ± SD).

The eleven organic acids detected in the recovered root exudates of Durum wheat during the 12-week growing period were lactic, acetic, propionic, butyric, pyruvic, valeric, glutaric, succinic, tartaric, maleic, and oxalic acids (Figure 6-18). The dominating organic acids in both treatments were lactic, acetic, butyric, and oxalic acids. The standard deviations for the minor organic acids were considerable indicating the variability of their productions during the growing season. For example, the propionic acid varied from n.d. to 0.17×10^{-3} and from n.d. to 0.33×10^{-3} moles COO^- per pot per collection period for the nutrient solution and the biosolids treatments, respectively. (Limit of detection is 0.01×10^{-3} M COO^- per pot per collection period.) In these cases, the calculated coefficients of variation were >100%.

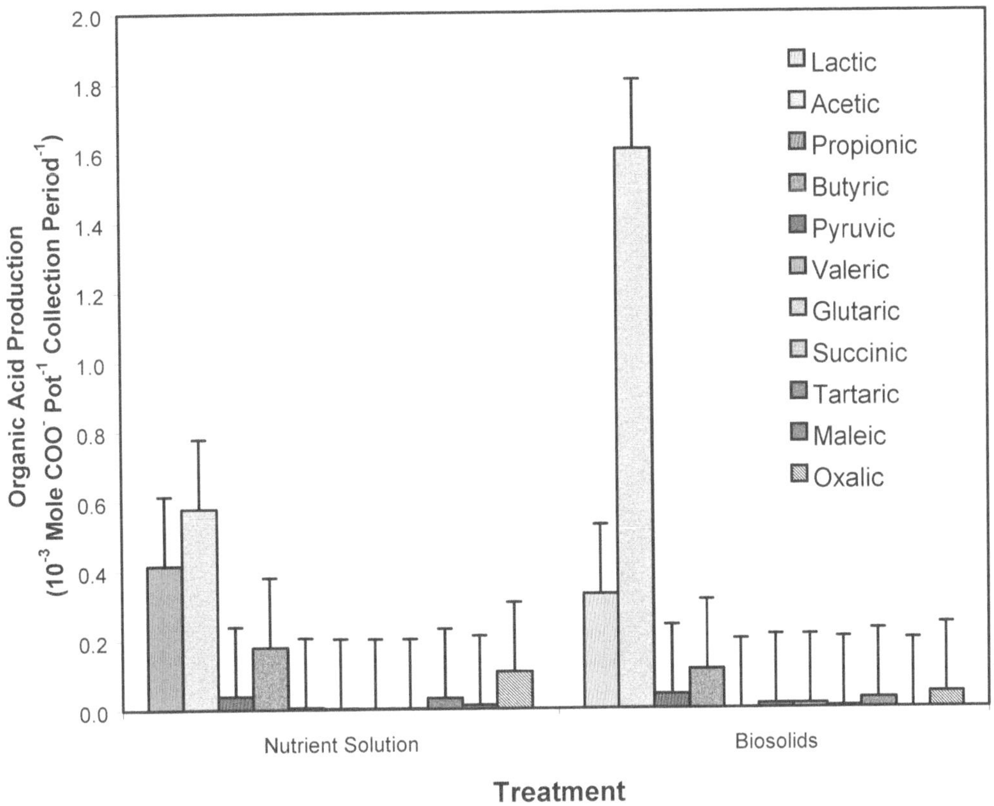

Figure 6-18. Organic Acids in the Rhizosphere of Durum Wheat Grown for 12 Weeks on Standard Plant Nutrient Solution- and Biosolids-Treated Media (Average ±SD).

Under the same treatment, the average organic acid productions in the rhizosphere for the three varieties of wheat were comparable. For Chinese spring, Yecora Rojo, and Durum wheat, amounts of organic acids produced were 1.1, 1.2 and 1.4 x 10^{-3} mole COO^- per pot per collection period for the standard nutrient solution and 2.2, 2.1, 2.2 x 10^{-3} mole COO^- per pot per collection period for the biosolids treatment. The patterns of organic acid distribution in the exudates of the Chinese spring and Yecora Rojo were similar to that of the previously described Durum wheat. Under the standard plant nutrient treatment, lactic, acetic, butyric, and oxalic acids accounted for >90% of the recovered organic acids (Figure 6-19). Under the biosolids treatment, acetic acid was the dominating component, accounting for 64-73% of the total recovered organic acids (Figure 6-20).

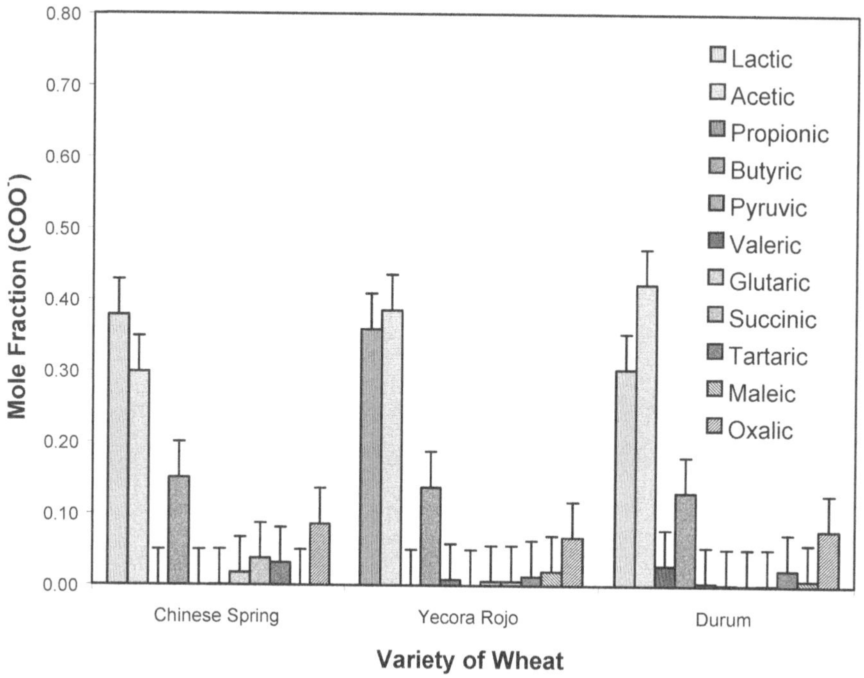

Figure 6-19. Organic Acids Distribution for Wheat Grown on Standard Plant Nutrient Solution Media (Average ± SD).

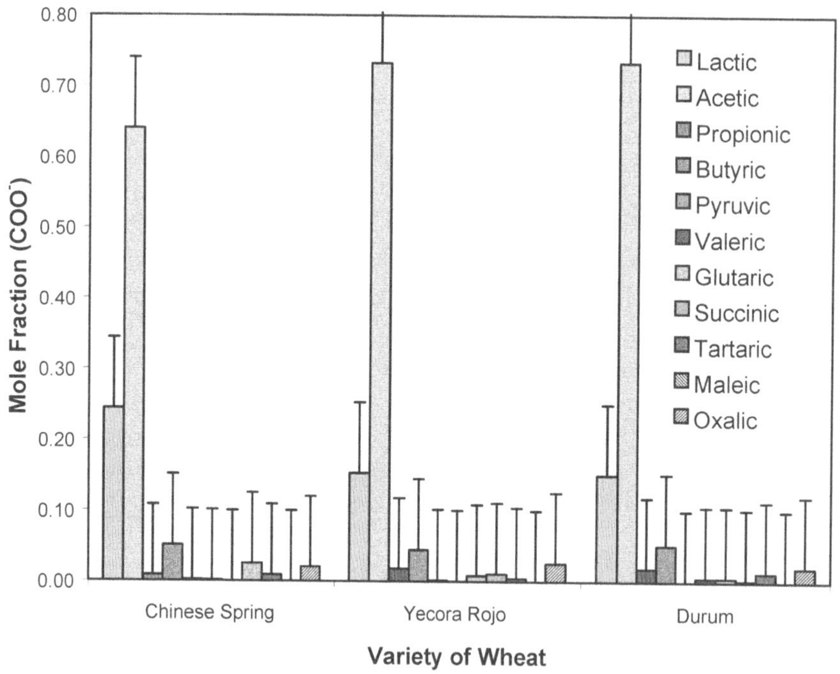

Figure 6-20. Organic Acids Distribution for Wheat Grown on Biosolids-Treated Media (Average ± SD).

6.6 Organic Acid in Root Exudates of Corn Grown on Five Different Biosolids

The average production of organic acids in the blank biosolids-treated culturing media and rhizosphere of corn grown on biosolids-treated media varied from 0.5–1.2 x 10^{-3} and 0.9–1.3 x 10^{-3} moles COO^- per pot per collection period, respectively. Considering the variability of the measurements, there was no significant difference in organic acid production between the blank media and the corresponding planted media (Figures 6-21 and 6-22). Judging from the patterns illustrated in Figures 6-22 and 6-23, there appeared to be significant differences between the organic acid composition of blanks and the planted media. In all cases, the major components in the root exudates were lactic, acetic, propionic, butyric, and oxalic acids. Other acids found were glutaric and succinic acids. The concentrations of pyruvic and tartaric acids were negligible. Butyric acid and acetic acid were the dominant components in the blank media and the planted media, respectively. The organic acid compositions of root exudates of various biosolids treatments did not vary significantly.

The relative abundances of organic acids in media treated with biosolids from different sources were similar (Figure 6-25). Lactic, acetic, propionic, butyric and oxalic acids always were present. Acetic, propionic, and butyric acids were the most abundant, accounting for 0.3 to > 0.5, 0.15 to 0.2, and 0.2 to 0.25 mole fraction COO^-, respectively. Lactic and oxalic acids averaged between 0.05-0.1 mole fractions of the total COO^- each. Smaller and varying amounts of pyruvic, glutaric, and succinic acids also were found (usually 0.01-0.05 mole fractions of COO^-). Citric acid was detected on several occasions. In all cases, it amounted to << 0.01 mole fraction of the total COO^- found in the biosolids-treated media.

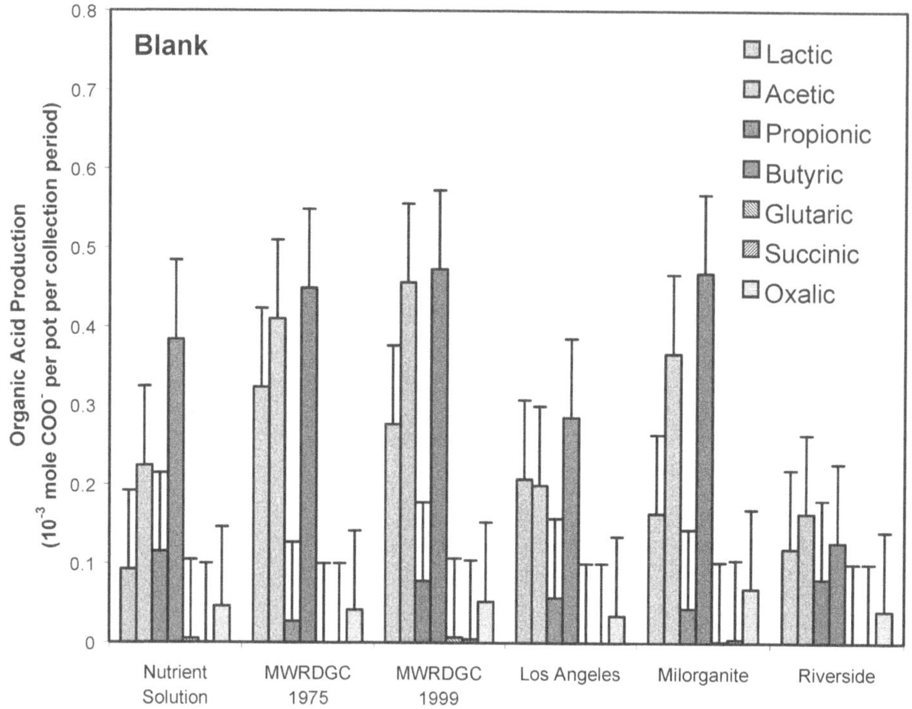

Figure 6-21. Organic Acids Recovered from the Without-Plant Blank of the Biosolids-Treated Media (Average ±SD).

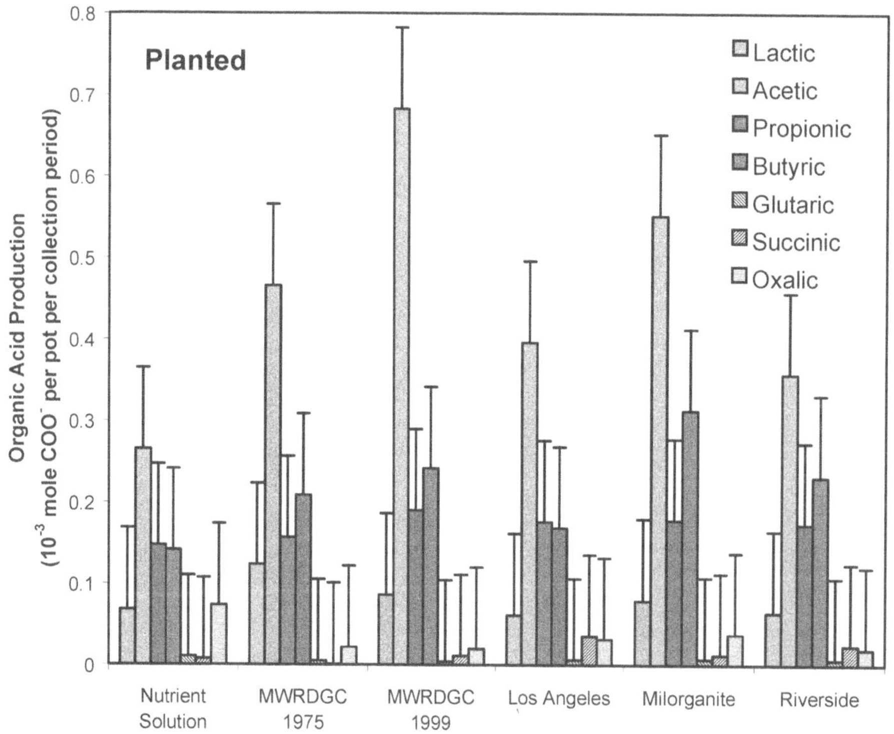

Figure 6-22. Organic Acids in the Rhizosphere of Corn Grown on Five Biosolids-Treated Media (Average ±SD).

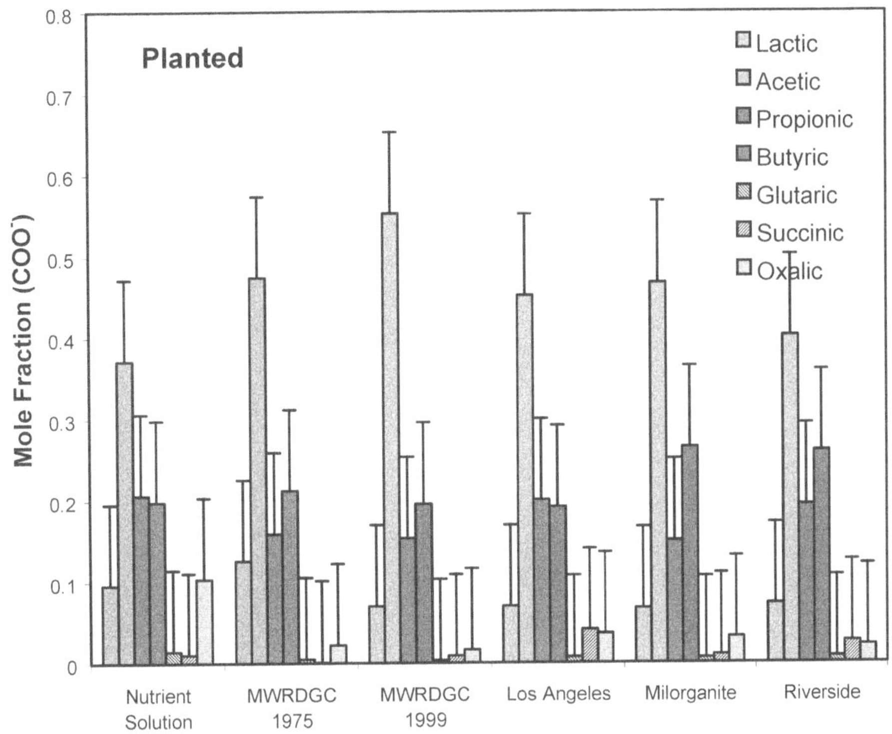

Figure 6-23. Composition of Organic Acids in Without-Plant Blank Biosolids-Treated Media (Average ± SD).

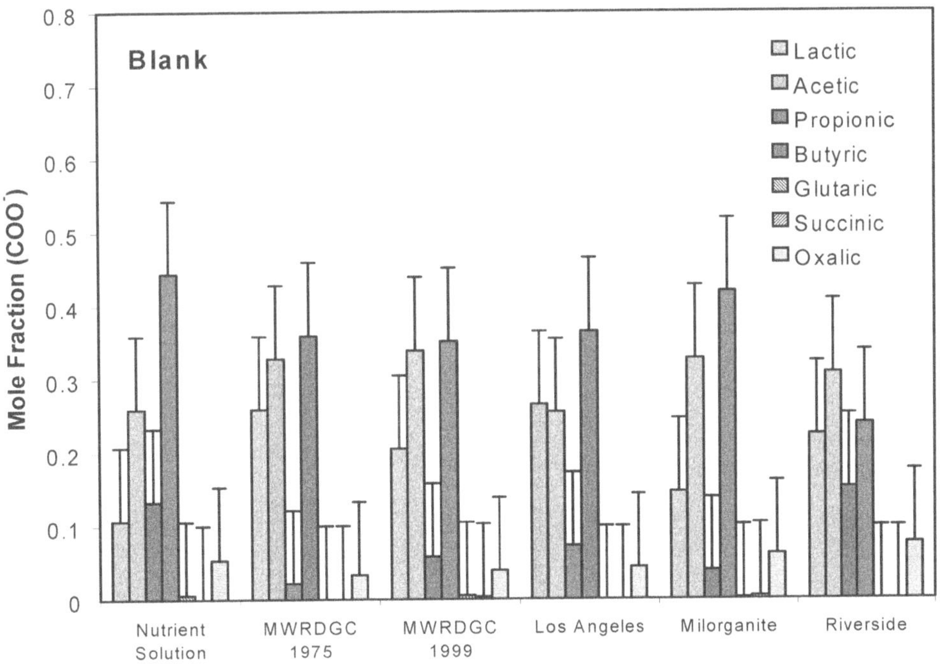

Figure 6-24. Composition of Organic Acids in the Rhizosphere of Corn Grown on Five Biosolids-Treated Media (Average ± SD).

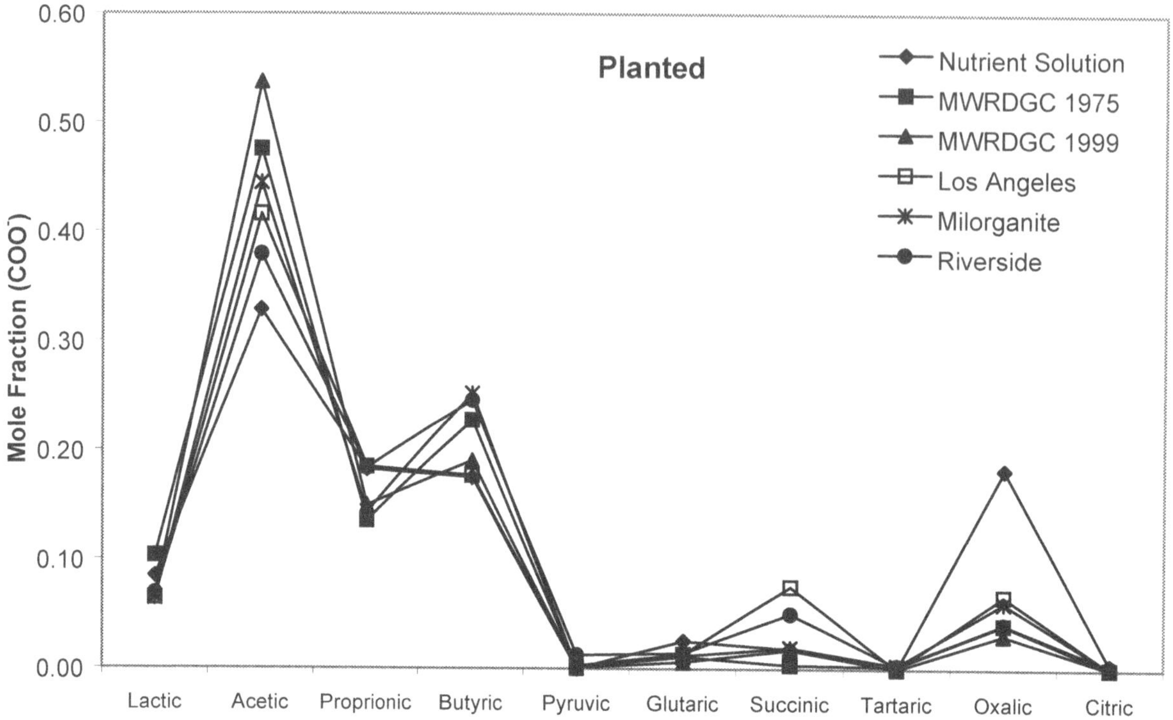

Figure 6-25. Relative Abundance of Organic Acids Recovered from Media Treated with Biosolids from Five Sources.

6.7 Summary

Experiments were conducted in a greenhouse, using sand media and biosolids-treated sand media that were irrigated with standard plant nutrient solutions, to investigate effects of biosolids amendments, plant species and varieties, and length of growing period on production and composition of organic acids in the rhizosphere. Findings are summarized as follows:

- Lactic, acetic, propionic, butyric, glutaric, maleic, valeric, succinic, tartaric, oxalic, and citric acids all were found in the rhizosphere. Among them, the lactic, acetic, butyric, and oxalic acids were always present and invariably the most abundant. They usually accounted for 60-80% of the total organic acids recovered. Other organic acids were found some of the times and in considerably less quantities.
- The amounts of organic acids recovered from the rhizosphere were dependent on the treatments. Plants and biosolids enhanced the organic acids production in the rhizosphere. In descending orders, amounts of organic acids recovered from rhizosphere of (planted and biosolids-treated media) > (planted and nutrient solution only media) > (without plant biosolids-treated blank media) > (without plant nutrient solution blank media).
- For the same treatment, the total amounts of organic acids recovered did not change significantly with the length of the plant growing period, plant species, plant varieties, or the source of biosolids.

- It appeared that the organic acids in the rhizosphere were produced through microbial metabolism. During the growing season of plants, the compositions dynamically changed depending on the availability of the substrate and the environmental conditions.

CHAPTER 7.0

DISSOLUTION OF BIOSOLIDS-BORNE METALS

Organic acids are preferable chemical reagents to extract metal contaminants from soils because the extraction can be performed at mildly acidic conditions (pH 3 to 5) and they are biodegradable (Veeken and Hamelers, 1999). The organic acids found in root exudates—such as citric, oxalic, tartaric, and acetic acids—are capable of forming complexes with Cu, Zn, Pb, and Co ions in solution (Rovira, 1969; Merckx et al., 1986; Inskeep and Comfort, 1986; Mench et al., 1988a; Cakmak and Marschner, 1988a, 1988b) and enhance their mobilization and uptake by plants (Treeby et al., 1989; Awad and Romheld, 2000). Romheld and Awad (2000) showed that plants exposed to low Fe nutrient media in metal-contaminated soil had a higher uptake of Fe, Zn, Ni, and Cd (up to 200%) than control plants that were exposed to adequate Fe nutrient media. Various organic acids and/or their salts were tested against two metal-chelating agents (EDTA and DTPA) for their potential effects in the remediation of a loam and a sandy clay loam soil polluted by heavy metals (Wasay et al., 1998). Experimental evidence showed that organic acids added to humic macromolecules induced the release of adsorbed metal ions (Nardi et al., 1986; Piccolo et al., 1992; Nardi et al., 1997). In near-neutral and neutral aqueous solution, organic acids are readily dissociated. The negatively charged organic acid ligands are capable of forming complexes with metals (e.g., Mn, Fe, and Zn) in solution, increasing their availability to plants (Jauregui and Reisenauer,1982; Dinkelaker et al.,1989; Rengel and Romheld, 2000).

The extent of complexation depends on the characteristics of organic acids involved (number and proximity of carboxylic groups), their concentrations, type of metal, and the pH of the soil. Among the frequently found organic acids in the rhizosphere, those with only one carboxyl group, such as acetic, formic, and lactic acids, have lesser metal complexing ability than maleic and oxalic acids, which have high affinity for complexing metals (Pohlman and McColl, 1986; Jones and Kochian, 1996). The ability of organic acids to complex metals also is dependent on pH (Mench and Martin, 1991). For instance, the complexation of Fe by malic and oxalic acids is highly dependent upon soil pH, with little or no complexation at pH > 7.0. In addition, malic and oxalic acids have the tendency to precipitate in the presence of Ca^{2+}, thus reducing their potential to complex with other metals. In general, in comparison to rainwater alone, organic acids are able to increase mineral dissolution rates by two to four times. The extent of the chemical reactions, however, is dependent upon mineral type, pH, and organic acid type (Pohlman and McColl, 1986; Jones and Kochian, 1996). Jones and Kochian (1996) reported that the presence of organic acids increases the dissolution of Fe and Al oxyhydroxides.

Organic acids may be adsorbed onto the hydroxyaluminum-montmorillonite (HyA-Mt) complex (Cambier and Sposito, 1991; Janssen et al., 1997). Cambier and Sposito (1991) concluded that the HyA-Mt complex was stable at $4 < pH < 5.5$, and only external HyA polycations could react with citrate. Janssen et al. (1997) noted that citrate did not appear to be adsorbed on the Al-OH groups of the HyA-Mt complex, but rather to the edge of the clay. Sakurai and Huang (1995) studied the effect of oxalate on the adsorption of Cd by montmorillonite (Mt) and HyA-Mt complex at pH 5. The reaction was very rapid and virtually completed within 10 minutes. The presence of oxalate markedly interfered with the Cd adsorption on clays, especially on the montmorillonite. Taniguchi et al. (2000) investigated the adsorption phenomena of Cd on the hydroxyaluminosilicate (HAS)-montmorillonite (Mt) and HyA-Mt complexes as influenced by oxalate and citrate. They concluded that the optimal concentrations of oxalate and citrate for Cd adsorption depended on the form of Cd ions in the solution.

Eaton (1936) observed that plant roots absorbed P and Fe from "water insoluble" P- and Fe-containing minerals present in the growth medium. Jenny and Overstreet (1938) proposed that reactions between root and solid phase minerals that were in direct contact or in close proximity would facilitate plant absorption of sparingly soluble mineral nutrients in soils. Recent studies have demonstrated that root exudates, which contain organic acids, play a significant role in the solubilization of metals in soil (Mench and Martin, 1991; Krishnamurti et al., 1997). In biosolids, metals are present almost entirely in solid phases. Upon being land applied, biosolids-borne metals remain largely in their original solid phases in biosolids-treated soils (Candalaria and Chang, 1997). Mobilization of metals by organic acids in root exudates may be the most significant pathway through which plant absorption of metals from biosolids-treated soils occurs. The total amount of bioavailable metals in biosolids-treated soils can be estimated by extracting the soils using organic acids found in the rhizosphere of plants grown on biosolids-treated medium. It follows that the rate at which metal becomes available to plants would be in proportion to the rate of metal dissolution into the organic acid mixture.

Based on the composition of organic acids in the rhizosphere (Chapter 6.0), synthetic organic acid mixtures were formulated and used as a surrogate for the actual organic acids in root exudates. Synthetic mixtures were used to study dissolution of metals in biosolids-treated soils, including the effects of the concentrations of the organic acid mixtures on metals solubility. Synthetic organic acid mixture-specific metal solubility and dissolution rate constants were determined for biosolids-treated soils.

The Romona soils (fine-loamy, mixed, themic Typic Haploxeralf) that were collected from a long-term field experiment on land application of biosolids at University of California, Riverside were selected. Briefly, a biosolids land application field experiment was set up in 1976 to investigate the metal accumulation in plant tissue and soils. The MWRDGC 1975 biosolids was applied at rates of 0 (control), 22.5 Mg ha^{-1} yr^{-1}, 45 Mg ha^{-1} yr^{-1}, 90 Mg ha^{-1} yr^{-1}, and 1080 Mg ha^{-1} yr^{-1} from 1976 through 1981. Following each application, varieties of vegetables were grown and harvested to measure the plant uptake of metals from biosolids-treated soils. At the termination of applications, the cumulative biosolids inputs amounted to 0 (control), 135 Mg ha^{-1}, 270 Mg ha^{-1}, 540 Mg ha^{-1}, and 1080 Mg ha^{-1}. The cultivation continued on the experimental plots until 1991 when the biosolids-treated soils were collected and archived for future experiments (Chang et al., 1997). Three of the soils were used for metal dissolution experiments described in the following sections, namely, the control, 135 Mg ha^{-1} and 1080 Mg ha^{-1}.

7.1 Formulation of Synthetic Root Exudates

Organic acids in the rhizosphere are difficult to collect because the volume produced is small and the components of the root exudates are readily biodegradable. To evaluate the ability of organic acids in the rhizosphere to solubilize soil metals, a large amount of root exudates must be used. Because of the difficulty in collecting and preserving large volumes of root exudates, it is imperative that synthetic root exudates be formulated for these studies.

It is not likely that the complete chemical matrix of the root exudates can be precisely mimicked. The synthetic root exudates, however, should contain the primary chemical components responsible for metal complex formation and these components should be in the concentration and pH ranges commonly observed in the rhizosphere. In addition, the synthetic root exudates should be prepared with the same background chemical matrix as the soil solution. In this manner, the synthetic organic acid mixtures would exhibit comparable ability to react with metals as those actually recovered from the rhizosphere.

A series of experiments was conducted to test effects of various synthetic organic acid mixture compositions and the concentrations in solubilizing metals in biosolids-treated soils containing a variety of metal concentrations. The amounts of metals extracted by the synthetic organic acid mixtures representing compositions of organic acids recovered from the rhizosphere of biosolids-treated medium and the standard sand medium may be illustrated by the Cd extractions summarized in Table 7-1. The differences of metals extracted due to the compositions representing organic acids at 2, 4, 8, 12, and 16 weeks of plant growth were not significantly at $p < 0.05$ and the differences due to organic acids of standard and biosolids-treated media were not significantly different at $p < 0.05$. The trends were similar for other metals extracted. The data were then pooled to test the differences in metal extractions due to the concentrations of synthetic organic acid mixtures. They were significantly different at $p < 0.05$.

Table 7-1. Amounts of Cd Extracted by Synthetic Organic Acid Mixtures of Various Compositions and Concentrations.

Organic Acid Concentration	Organic Acid Composition[a]	Cd Concentration (mg kg^{-1})[b]	
		Biosolids	Nutrient Solution
0.1 M	2-week	2.11 Ba	1.86 Ba
	4-week	2.36 Aa	2.58 Aa
	8-week	1.98 Ba	1.63 Ba
	12-week	1.79 Ba	1.60 Ba
	16-week	1.71 Ba	1.62 Ba
0.01 M	2-week	0.39 Aa	0.31 Aa
	4-week	0.41 Aa	0.43 Aa
	8-week	0.44 Aa	0.35 Aa
	12-week	0.47 Aa	0.37 Aa
	16-week	0.51 Aa	0.43 Aa
0.001 M	2-week	0.29 Aa	0.20 Aa
	4-week	0.32 Aa	0.24 Aa
	8-week	0.36 Aa	0.32 Aa
	12-week	0.29 Aa	0.23 Aa
	16-week	0.30 Aa	0.29 Aa
13.5 mM $Ca(NO_3)_2$	Not Applicable	0.10	

[a]Compositions corresponding to that of organic acids recovered from nutrient solution only and biosolids-treated rhizosphere of corn at Weeks 2, 4, 8, 12, and 16 of the growth period.

[b]Letters following the value indicate statistical significance, where values followed by the same letter were not statistically significant at $p<0.05$.

Values reported in Table 7-1 represent means of four replicates. The differences of Cd concentrations among the organic acid compositions of growth periods were tested by one-way ANOVA. In each column of each organic acid mixture concentration, values followed by the same upper case letter were not significantly different at $p < 0.05$. The differences of the Cd concentrations between standard and biosolids treatments were tested by the Student-Newman-Keuls test. Each pair of Cd concentrations followed by the same lower case letter was not significantly different at $p < 0.05$. Based on the results, the synthetic root exudates were formulated as described in the following section.

7.1.1 Composition of Organic Acids

The organic acid compositions of the root exudates recovered from plants grown for various lengths of time (Section 6.3) and from different plant species (Section 6.4) did not vary significantly. Further, the organic acids composition of exudates was not significantly different among the various biosolids treatments tested (Section 6.6). Thus, the synthetic exudate was formulated based on the average composition of organic acids recovered from the rhizosphere of corn grown on biosolids-treated medium for 16 weeks (Table 7-2).

Table 7-2. Composition of Synthetic Exudate (Based on Organic Acids Composition in the Rhizosphere of Corn Grown on Biosolids-Treated Soil).

Organic Acid	Molecular Weight	Formula	Mole Fraction (COO^-)	Number of COOH Functional Groups
Acetic	60.05	$C_2H_4O_2$	0.287	1
Butyric	88.11	$C_4H_8O_2$	0.209	1
Glutaric	132.12	$C_5H_8O_4$	0.004	2
Lactic	90.08	$C_3H_6O_3$	0.366	1
Maleic	116.07	$C_4H_4O_4$	0.042	2
Oxalic	90.04	$C_2H_2O_4$	0.043	2
Propionic	74.08	$C_3H_6O_2$	0.010	1
Pyruvic	88.06	$C_3H_4O_3$	0.0004	1
Succinic	118.09	$C_4H_6O_4$	0.006	2
Tartaric	150.09	$C_4H_6O_6$	0.032	2
Valeric	102.13	$C_5H_{10}O_2$	0.001	1

7.1.2 pH of the Rhizosphere Soils

The rhizosphere is the portion of the soil that is next to the root of living plants and is subject to the influence of chemicals excreted by roots and the microbial community supported by the root exudation. The rhizosphere domain varies with the plant species and age and morphology of roots. Generally, it extends 1-5 mm outward from the interface of the root and soil. However, pH measurement in this reach of soil is complicated. Instead of directly measuring the pH of the rhizosphere, it was deduced and estimated from data found in the literature (Table 3.2). It was apparent that the pH at or near the root of living plants was altered by the root exudation. It varied from pH 4-5 at the soil root interface and gradually varied to the level comparable to the pH of the bulk soils over approximately 5 mm of distance. A pH of 4.8 was chosen for the synthetic root exudates based on effects of pH on the dissolution of metals in the biosolids-treated soils, as summarized in Section 7.2.

7.1.3 Concentration of Organic Acids

Table 7-3 summarizes the amounts of metals extracted by synthetic exudates at concentrations of 0.001, 0.01, and 0.1 M. The experimental results indicate that the concentration of the synthetic exudate significantly affects the amounts of metals extracted. The amounts of metals extracted by the synthetic organic acid mixtures were considerably greater than that of the neutral electrolyte solution, 13.5 mM of $Ca(NO_3)_2$, which was used to represent the background chemical matrix of the soil solution.

As the root exudates were recovered from the rhizosphere with two liters of water (see Chapter 6.0), their soil solution concentrations must be estimated by dividing the amounts recovered (mass) with the soil moisture contents (volume). Based on the estimated moisture content of the rhizosphere (the actual moisture contents fluctuated and were not measured) and the average volume of organic acids recovered from the rhizosphere in the experiments described in Chapter 6.0, the concentrations of organic acids in the rhizosphere were estimated to be from 0.005 to 0.013 M (Table 7-4). As the organic acids are closely associated with the root exudates, their concentrations would be expected to be considerably higher near the root and soil interface and could approach 0.1 M.

Table 7-3. Metal Extracted from Control and Biosolids-Treated Romona Soil by Synthetic Exudate at Three Concentrations.

A. Romona Soil – Control

Element	Metals Extracted (mg kg^{-1})			
	13.5 mM $Ca(NO_3)_2$	Synthetic Exudate[a]		
		0.001 M	0.01 M	0.1 M
Cd	0.0008	0.0016	0.012	0.04
Cr	0.001	0.04	0.08	0.50
Cu	0.12	0.20	0.61	3.35
Ni	0.03	0.08	0.72	1.71
Pb	n.d.[b]	n.d.	n.d.	n.d.
Zn	0.18	0.39	3.2	8.1

B. Romona Soil - Biosolids-Treated (135 Mg ha^{-1})

Element	Metals Extracted (mg kg^{-1})			
	13.5 mM $Ca(NO_3)_2$	Synthetic Exudate		
		0.001 M	0.01 M	0.1 M
Cd	0.023	0.12	0.29	1.46
Cr	0.01	0.29	0.80	3.82
Cu	0.3	1.37	3.55	14.2
Ni	0.2	0.97	2.33	11.8
Pb	n.d.	0.07	0.15	1.22
Zn	1.32	6.23	40	155

C. Romona Soil - Biosolids-Treated (1,080 Mg ha^{-1})

Element	Metals Extracted (mg kg^{-1})			
	13.5 mM $Ca(NO_3)_2$	Synthetic Exudate		
		0.001 M	0.01 M	0.1 M
Cd	0.21	0.53	2.11	3.58
Cr	0.11	1.55	3.81	9.72
Cu	0.82	4.44	19	38
Ni	0.51	4.48	17.3	30.3
Pb	0.03	0.52	1.35	4.08
Zn	3.51	36	168	354

[a]Concentrations of the synthetic organic acid mixtures were 0.001, 0.01, and 0.1 M COO^-.
[b]Some of the observations were below detection limits of the AAS for Pb of 0.001 mg kg^{-1}.

Table 7-4. Estimated Solution Concentrations (16 Week Average) of Organic Acids in the Rhizosphere of Corn.

Treatment	Estimated Organic Acid Concentration (10^{-3} M)[a]	
	Standard Media	Biosolids-treated Media
Blank	2.05 (0.63)	3.41 (0.87)
Planted	5.23 (1.21)	12.9 (2.04)

[a]Values represent means ± SD of 20 replicates, based on four replicates at each of 2, 4, 8, 12, and 16 weeks of the experiment growing corn from seeding to maturity in hydroponics with standard and biosolids-treated sand media.

7.2 Effects of pH on the Dissolution of Metals in Biosolids-Treated Soils

The pH of the rhizosphere may vary from 4.0-8.0 and may affect the solubility of metals in the soil. The effects of pH on the dissolution of metals in biosolids-treated soils by the synthetic exudate were determined by evaluating metals releases at various pH values. A biosolids-treated Romona soil (135 Mg ha^{-1}) was extracted by 0.01 M synthetic exudate and 13.5 mM $Ca(NO_3)_2$ solutions that were adjusted to pH 4.5, 5.5, 6.5, and 7.5.

In general, the amounts of metal extracted decreased with the increase of the pH and, at the same pH level, the synthetic exudate extracted more metals than the 13.5 mM $Ca(NO_3)_2$ electrolyte solution (Table 7-5). The extent of pH-induced changes in metal solubility was not substantial. Cd and Zn experienced by far the biggest decreases with increasing pH; Cd and Zn solubility declined by half over the tested pH range, with the greatest change occurring from pH 4.5–5.5 (Tables 7-5). Statistical tests indicated that metals extracted at pH 6.5 were not significantly different from the corresponding metals extracted at pH 7.5. Based on the outcome of this experiment, a pH of 4.8 was selected for the synthetic exudate extractions in subsequent experiments.

Table 7-5. Effects of pH on Metals Releases from Biosolids-Treated Romona Soil (135 Mg ha^{-1}) by 13.5 mM $Ca(NO_3)_2$ Solution and 0.01 M Synthetic Exudate.

Element	pH	Metals Extracted (mg kg^{-1})[a]	
		13.5 mM $Ca(NO_3)_2$	0.01 M Synthetic Exudate
Cd	4.5	0.04 (0.002) Aa	0.04 (0.001) Aa
	5.5	0.02 (0.001) Bb	0.03 (0.001) Ba
	6.5	0.01 (0.001) Cb	0.02 (0.001) Ca
	7.5	0.01 (0.001) Cb	0.02 (0.001) Ca
Cr	4.5	0.08 (0.004) Ab	0.10 (0.003) Aa
	5.5	0.07 (0.002) Bb	0.09 (0.004) Ba
	6.5	0.07 (0.001) Bb	0.08 (0.001) Ca
	7.5	0.07 (0.003) Bb	0.08 (0.001) Ca
Cu	4.5	1.3 (0.08) Ab	1.8 (0.09) Aa
	5.5	1.0 (0.01) Bb	1.6 (0.06) Ba
	6.5	0.9 (0.04) Cb	1.5 (0.03) Ca
	7.5	0.9 (0.02) Cb	1.4 (0.03) Da
Ni	4.5	0.9 (0.03) Ab	1.1 (0.04) Aa
	5.5	0.6 (0.04) Bb	0.9 (0.02) Ba
	6.5	0.6 (0.01) Bb	0.9 (0.03) Ba
	7.5	0.5 (0.01) Cb	0.8 (0.01) Ca
Zn	4.5	2.0 (0.04) Ab	2.3 (0.05) Aa
	5.5	1.3 (0.06) Bb	1.7 (0.03) Ba
	6.5	1.0 (0.03) Cb	1.5 (0.01) Ca
	7.5	0.5 (0.03) Db	1.1 (0.05) Da

[a]Values represent means and standard deviations (in parenthesis) of four replicates. The differences of metal concentrations among the pH values were tested by one-way ANOVA. In each column, values followed by the same upper case letter were not significantly different at $p < 0.05$. The differences of the metal concentrations between 13.5 mM $Ca(NO_3)_2$ and 0.01 M synthetic organic acid mixture were tested by the Student-Newman-Keuls test. In each row, values followed by the same lower case letter were not significantly different at $p < 0.05$.

7.3 Metal Solubility in Organic Acids Mixtures

Metals in the soils were not readily extractable by the neutral electrolyte solution blank that contained 13.5 mM $Ca(NO_3)_2$ and, in general, less than 0.25% of any metals were solubilized (Tables 7-6). For Pb, the concentrations in the extracts of the control soil were below the limits of quantification of atomic absorption spectroscopy (< 0.001 mg kg^{-1}). For the control soil, the total metal concentrations were considerably lower than those in biosolids-treated soils and the percent of metals extracted also was lower than those of the biosolids-treated soils.

Table 7-6. Percent of Metal Extracted from Control and Biosolids-Treated Romona Soil by Synthetic Exudate at Various Concentrations .

Element	Biosolids Treatment[a]	Metal Extracted (% of Total in Soil) [b]			
		$Ca(NO_3)_2$ Blank	Synthetic Exudate		
			0.001 M	0.01 M	0.1 M
Cd	Control	0.16 (0.01)	0.32 (0.01)	2.40 (0.01)	8.01 (0.01)
	135 Mg ha^{-1}	0.21 (0.01)	1.09 (0.01)	2.64 (0.04)	13.3 (0.05)
	1,080 Mg ha^{-1}	0.23 (0.01)	2.04 (0.03)	8.12 (0.10)	13.8 (0.16)
Cr	Control	0.003 (0.0001)	0.11 (0.002)	0.22 (0.01)	1.35 (0.03)
	135 Mg ha^{-1}	0.004 (0.0001)	0.12 (0.01)	0.33 (0.01)	1.57 (0.11)
	1,080 Mg ha^{-1}	0.019 (0.001)	0.26 (0.03)	0.64 (0.31)	1.63 (0.29)
Cu	Control	0.11 (0.001)	0.20 (0.01)	0.70 (0.02)	3.19 (0.03)
	135 Mg ha^{-1}	0.16 (0.001)	0.73 (0.05)	1.89 (0.10)	7.55 (0.33)
	1,080 Mg ha^{-1}	0.17 (0.001)	0.93 (0.14)	3.98 (0.46)	7.95 (1.35)
Ni	Control	0.13 (0.001)	0.33 (0.004)	2.20 (0.01)	7.10 (0.03)
	135 Mg ha^{-1}	0.23 (0.001)	1.10 (0.01)	2.65 (0.05)	13.4 (0.16)
	1,080 Mg ha^{-1}	0.24 (0.001)	2.08 (0.23)	8.05 (0.64)	14.1 (0.57)
Pb	Control	n.d.[c]	n.d.	n.d.	n.d.
	135 Mg ha^{-1}	n.d.	0.06 (0.01)	0.13 (0.05)	1.02 (0.16)
	1,080 Mg ha^{-1}	0.008(0.001)	0.13 (0.03)	0.34 (0.08)	1.03 (0.05)
Zn	Control	0.18(0.01)	0.41 (0.02)	3.37 (0.05)	8.50 (0.12)
	135 Mg ha^{-1}	0.24(0.01)	1.11 (0.03)	7.16 (0.19)	27.7 (1.61)
	1,080 Mg ha^{-1}	0.24(0.01)	2.46 (0.90)	11.5 (1.30)	24.2 (4.70)

[a]Obtained at Moreno Field Station of the University of California, Riverside, CA. From 1976 through 1981, composted biosolids were applied at dry weight rates of 0 (control), 22.5 and 180 Mg ha^{-1} yr^{-1}, respectively.
[b]Values represent means and standard deviations (in parenthesis).
[c] n.d, below detection limits of AAS for Pb (< 0.001 mg kg^{-1}).

In biosolids-treated soils, the amounts of metals extracted were proportional to the amounts present and the percentage of the total metals extracted from biosolids-treated soils increased with the concentration of the synthetic exudate. The percentage of extraction did not appear to change significantly with the biosolids loading when soils were extracted by the 0.1 M synthetic exudate. In general, Cd, Cu, Ni, and Zn were more readily extractable by the synthetic organic acid mixtures (Tables 7-6) than Cr and Pb. Metals extracted from biosolids-treated soils came predominantly from the biosolids. For instance, the amount of Cd extracted from control Romona soil by 13.5 mM $Ca(NO_3)_2$ was 0.0008 mg kg^{-1}, whereas the amount of Cd extracted from biosolids-treated soil by the same electrolyte solution was 0.023 mg kg^{-1} (Table 7-3). We calculated that, in this case, 97% of the Cd extracted from biosolids-treated soils was biosolids-borne.

7.4 Metal Dissolution Kinetics of Biosolids-Treated Soils in Synthetic Organic Acid Mixtures

7.4.1 Results of Equilibration Studies

The batch equilibrium method was used to study the kinetics of metal dissolution in biosolids-treated soils. The experimental procedures were similar to those previously described in the metal solubility study, with the exception that samples were equilibrated for time periods ranging from 15 minutes to 48 hours. When the biosolids-treated soils (1,080 Mg ha^{-1}) were equilibrated with 0.1 M synthetic exudate, the amounts of biosolids-borne metals solubilized in the synthetic organic acid mixtures increased with increasing equilibration time (Table 7-7). Reactions with organic acids at 0.001 and 0.01 M and soils with control and lower biosolids inputs yielded similar results.

Table 7-7. Metals Release Kinetics of Biosolids-Treated Romona Soil (1,080 Mg ha^{-1}) in Different Concentrations of Synthetic Exudate

0.1 M Synthetic Exudate

Element	Metal Concentration (mg kg^{-1})							
	15 min.	30 min.	1 hr.	2 hrs.	4 hrs.	8 hrs.	24 hrs.	48 hrs.
Cd	1.67	1.70	1.75	1.82	1.90	2.03	2.40	2.48
Cr	1.46	1.78	1.91	2.17	2.53	3.17	4.79	5.04
Cu	20.3	21.2	22.5	23.7	25.2	26.4	32.1	33.4
Ni	16.1	16.8	18.0	18.8	19.5	20.8	23.3	24.1
Pb	0.77	0.82	0.87	1.02	1.09	1.89	2.01	2.33
Zn	205	211	220	227	235	245	275	282

0.01 M Synthetic Exudate

Element	Metal Concentration (mg kg^{-1})							
	15 min.	30 min.	1 hr.	2 hrs.	4 hrs.	8 hrs.	24 hrs.	48 hrs.
Cd	1.16	1.21	1.24	1.28	1.35	1.49	1.73	1.75
Cr	0.85	0.95	1.03	1.16	1.39	1.67	2.49	2.57
Cu	10.7	11.1	11.7	12.1	12.8	13.8	16.8	17.5
Ni	9.80	10.2	10.9	11.4	12.1	12.5	14.5	14.8
Pb	0.34	0.36	0.38	0.45	0.54	0.67	0.99	1.03
Zn	104	108	111	115	123	129	145	148

0.001 M Synthetic Exudate

Element	Metal Concentration (mg kg^{-1})							
	15 min.	30 min.	1 hr.	2 hrs.	4 hrs.	8 hrs.	24 hrs.	48 hrs.
Cd	0.31	0.32	0.33	0.35	0.37	0.40	0.46	0.47
Cr	0.34	0.40	0.44	0.49	0.57	0.73	1.09	1.12
Cu	2.41	2.48	2.63	2.78	2.88	3.14	3.81	3.89
Ni	2.53	2.64	2.83	2.94	3.13	3.28	3.72	3.79
Pb	0.13	0.14	0.15	0.18	0.22	0.29	0.42	0.46
Zn	25	26	28	29	31	32	36	38

Table 7-8. Metals Release Kinetics of Biosolids-Treated Romona Soil (135 Mg ha^{-1}) in Different Concentrations of Synthetic Exudate.

0.1 M Synthetic Exudate

Element	Metal Concentration (mg kg^{-1})							
	15 min.	30 min.	1 hr.	2 hrs.	4 hrs.	8 hrs.	24 hrs.	48 hrs.
Cd	0.79	0.83	0.88	0.92	1.01	1.10	1.26	1.29
Cr	0.91	1.07	1.13	1.23	1.47	1.80	2.63	2.68
Cu	7.6	7.8	8.3	8.9	9.4	10.1	12.6	12.7
Ni	6.8	7.3	7.8	8.4	8.9	9.5	11.0	11.3
Pb	0.32	0.34	0.38	0.45	0.53	0.62	0.90	0.94
Zn	100	103	108	114	117	121	134	137

0.01 M Synthetic Exudate

Element	Metal Concentration (mg kg^{-1})							
	15 min.	30 min.	1 hr.	2 hrs.	4 hrs.	8 hrs.	24 hrs.	48 hrs.
Cd	0.18	0.19	0.20	0.21	0.23	0.25	0.29	0.30
Cr	0.23	0.27	0.30	0.34	0.38	0.46	0.69	0.71
Cu	2.21	2.35	2.49	2.63	2.80	2.93	3.59	3.66
Ni	1.57	1.67	1.80	1.93	2.01	2.13	2.52	2.57
Pb	0.06	0.07	0.08	0.09	0.11	0.13	0.19	0.20
Zn	21.4	22.6	23.3	24.8	25.7	27.1	30.7	31.0

0.001 M Synthetic Exudate

Element	Metal Concentration (mg kg^{-1})							
	15 min.	30 min.	1 hr.	2 hrs.	4 hrs.	8 hrs.	24 hrs.	48 hrs.
Cd	0.055	0.059	0.063	0.066	0.068	0.073	0.089	0.091
Cr	0.07	0.08	0.09	0.10	0.13	0.16	0.22	0.23
Cu	0.86	0.88	0.92	0.97	1.05	1.15	1.38	1.42
Ni	0.53	0.55	0.59	0.62	0.66	0.68	0.79	0.81
Pb	0.021	0.023	0.025	0.030	0.035	0.044	0.063	0.065
Zn	3.96	4.19	4.35	4.48	4.69	4.91	5.60	5.66

Table 7-9. Metals Release Kinetics of Control Romona Soil in Different Concentrations of Synthetic Exudate.

0.1 M Synthetic Exudate

Element	Metal Concentration (mg kg^{-1})							
	15 min.	30 min.	1 hr.	2 hrs.	4 hrs.	8 hrs.	24 hrs.	48 hrs.
Cd	0.020	0.021	0.022	0.023	0.025	0.027	0.031	0.032
Cr	0.11	0.13	0.14	0.16	0.20	0.23	0.35	0.36
Cu	1.97	2.03	2.12	2.21	2.34	2.51	3.11	3.17
Ni	1.07	1.12	1.18	1.24	1.33	1.42	1.64	1.67
Pb	0.014	0.016	0.017	0.019	0.024	0.030	0.042	0.044
Zn	5.3	5.6	6.0	6.2	6.5	6.8	7.8	7.9

0.01 M Synthetic Exudate

Element	Metal Concentration (mg kg^{-1})							
	15 min.	30 min.	1 hr.	2 hrs.	4 hrs.	8 hrs.	24 hrs.	48 hrs.
Cd	0.0084	0.0088	0.0091	0.0097	0.0102	0.0115	0.0137	0.014
Cr	0.015	0.018	0.020	0.023	0.028	0.035	0.052	0.053
Cu	0.35	0.36	0.39	0.41	0.44	0.47	0.56	0.58
Ni	0.39	0.42	0.47	0.48	0.51	0.53	0.63	0.64
Pb	n.d.[a]	n.d.	n.d.	n.d.	n.d.	n.d.	n.d.	n.d.
Zn	1.95	2.06	2.17	2.26	2.34	2.44	2.77	2.79

0.001 M Synthetic Exudate

Element	Metal Concentration (mg kg^{-1})							
	15 min.	30 min.	1 hr.	2 hrs.	4 hrs.	8 hrs.	24 hrs.	48 hrs.
Cd	0.0013	0.0014	0.0015	0.0015	0.0016	0.0017	0.002	0.002
Cr	0.010	0.012	0.013	0.016	0.019	0.023	0.034	0.035
Cu	0.1071	0.1139	0.1207	0.1275	0.1343	0.1433	0.170	0.177
Ni	0.0603	0.063	0.0675	0.0702	0.0737	0.0773	0.0891	0.090
Pb	n.d.	n.d.	n.d.	n.d.	n.d.	n.d.	n.d.	n.d.
Zn	0.30	0.32	0.34	0.35	0.37	0.38	0.44	0.45

[a]n.d., concentraton was below detection limits of the AAS for Pb (< 0.001 mg kg^{-1}).

The percentages of total biosolids-borne metals extracted were 13.8, 1.63, 7.95, 14.1, 1.03, and 24.2% for Cd, Cr, Cu, Ni, Pb, and Zn, respectively. If organic acids are responsible for converting metals in the solid phases into plant-available forms, the amounts and rates of metals solubilized would be indicative of the metals' availability to plants.

The metal dissolution kinetics data were plotted as the fraction of the total dissolved metals (C_{48}, based on 48 hours equilibration) that was found at time t (C_t, amount dissolved at t) (Figures 7-1 to 7-3). The patterns for dissolution in the synthetic organic acid mixtures were essentially the same for all metals. When the biosolids-treated soils were equilibrated with the 0.1 M synthetic exudate, there appeared to be an immediate and rapid release of metals from the soil. The amounts of Cd, Cu, Ni, and Zn solubilized in the first 15 minutes ($C_{0.25}$) accounted for approximately 60-75% of the total dissolved metals. For Cr and Pb, approximately eight hours was needed to dissolve 60-70% of the total soluble metals. After 15 minutes, the dissolution of metals slowly reached the steady state over a period of 48 hours. When the concentrations of the synthetic exudates were 0.001 M and 0.01 M, the dissolution behavior of metals in biosolids-treated soils exhibited the same pattern as those seen with 0.1 M synthetic exudates, although the amounts of metals extracted were considerably less. The percentages of the total soluble metals that entered into the solution phase in the first 15 minutes, however, were essentially identical for 0.001, 0.01, and 0.1 M synthetic organic acid mixtures (Table 7-10).

Table 7-10. Metals of Standard and Biosolids-Treated Soils Solubilized by Synthetic Organic Acid Mixtures in the First 15 Minutes of the Equilibration.

Biosolids Treatment[a]	Synthetic Exudate	Metals Dissolved in 15 Minutes (% of Total Dissolved)					
		Cd	Cr	Cu	Ni	Pb	Zn
Control	0.001 M	63	29	61	67	n.d.[b]	67
	0.01 M	60	28	60	61	n.d.	70
	0.1 M	63	31	62	64	32	67
135 Mg ha^{-1}	0.001 M	60	30	61	66	32	70
	0.01 M	60	32	60	61	30	69
	0.1 M	62	34	60	60	34	73
1,080 Mg ha^{-1}	0.001 M	66	30	62	67	28	66
	0.01 M	66	33	61	66	33	70
	0.1 M	67	29	61	67	33	73

[a]Obtained at Moreno Field Station of the University of California, Riverside, CA. From 1976 through 1981 composted biosolids were applied at dry weight rates of 0 (control), 22.5 and 180 Mg ha^{-1} yr^{-1}, respectively.
[b]n.d., concentration was below detection limits of the AAS for Pb (< 0.001 mg kg^{-1}).

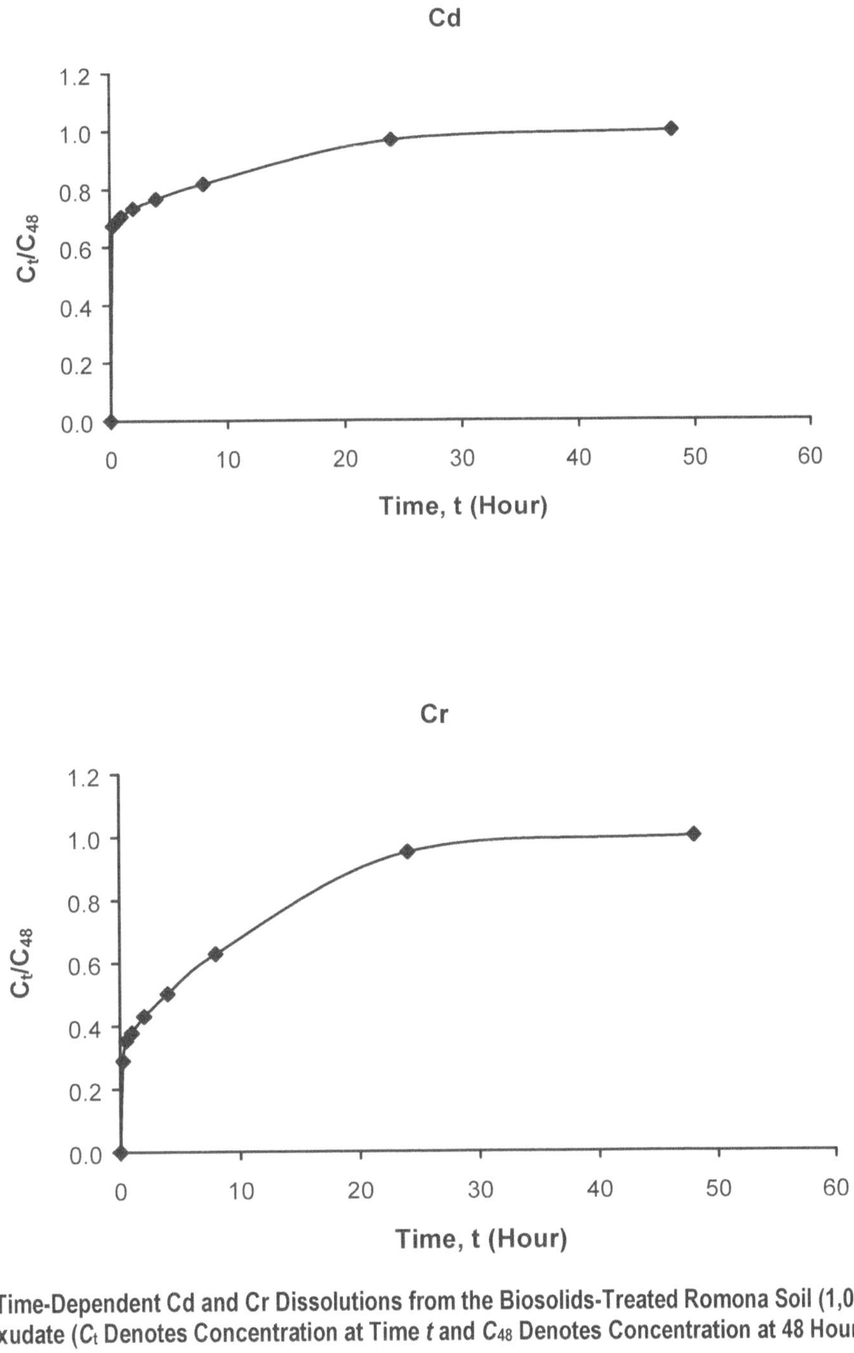

Figure 7-1. Time-Dependent Cd and Cr Dissolutions from the Biosolids-Treated Romona Soil (1,080 Mg ha^{-1}) in 0.1M Synthetic Exudate (C_t Denotes Concentration at Time *t* and C_{48} Denotes Concentration at 48 Hours).

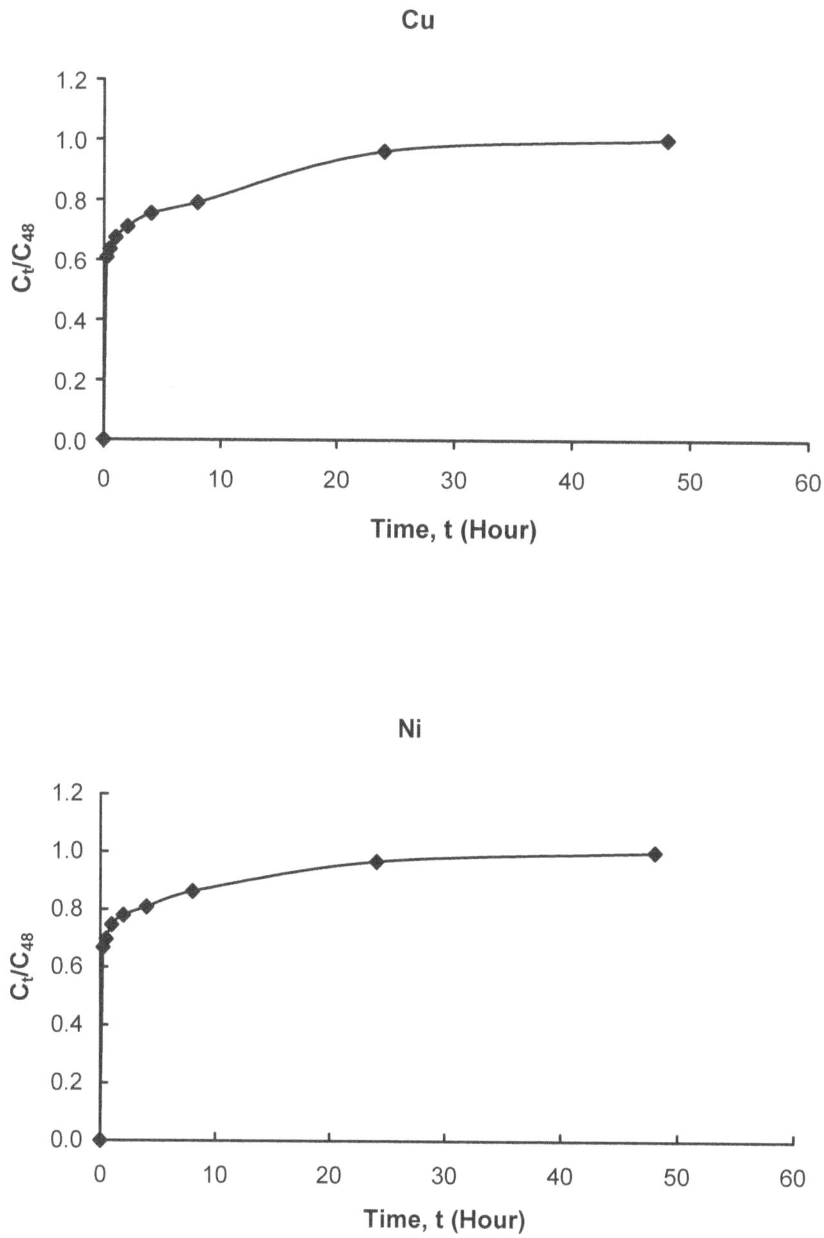

Figure 7-2. Time-dependent Cu and Ni Dissolutions from the Biosolids-Treated Romona Soil (1,080 Mg ha^{-1}) in 0.1M Synthetic Exudate (C_t Denotes Concentration at Time t and C_{48} Denotes Concentration at 48 Hours).

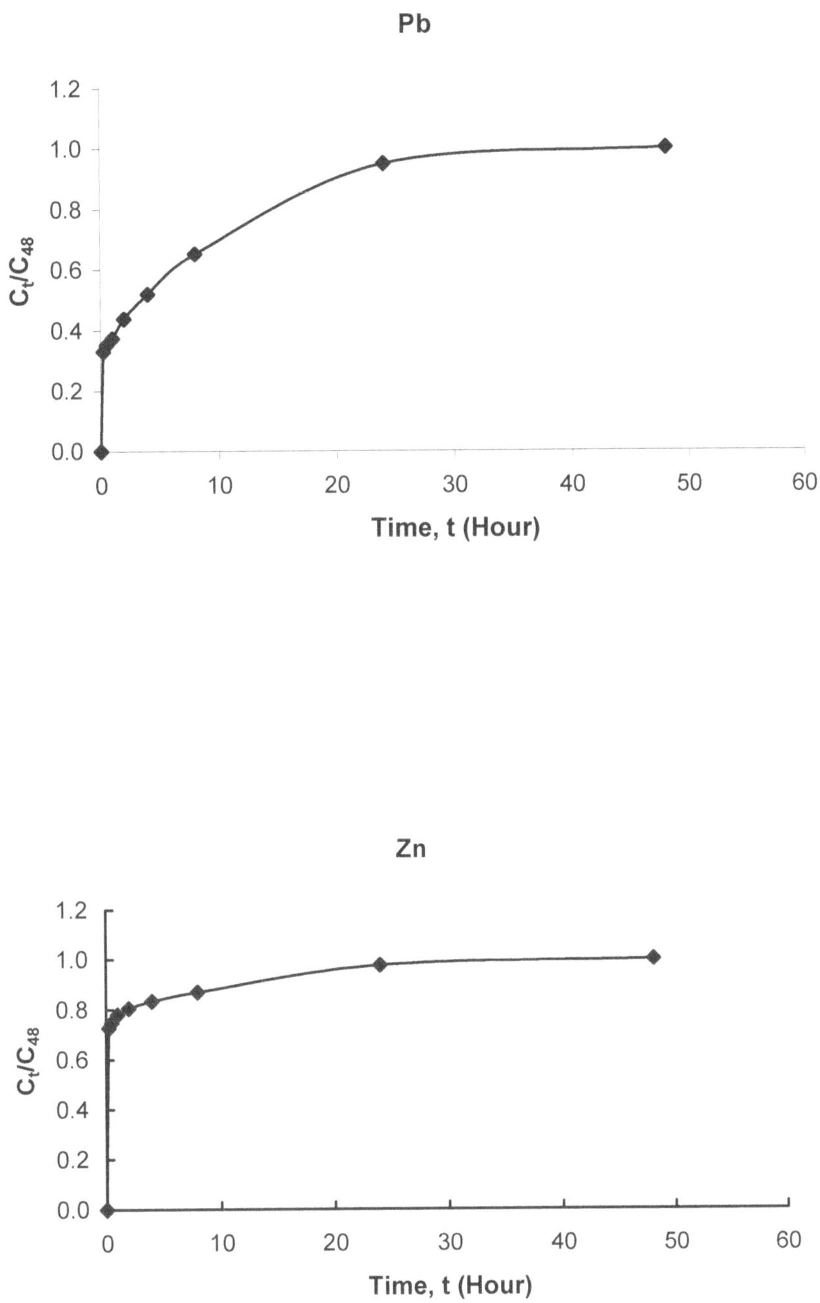

Figure 7-3. Time-dependent Pb and Zn Dissolutions from the Biosolids-Treated Romona Soil (1,080 Mg ha^{-1}) in 0.1M Synthetic Exudate (C_t Denotes Concentration at Time t and C_{48} Denotes Concentration at 48 Hours).

7.4.2 Bi-continuum Model

A variety of chemical reactions occur in soils and reactions often take place simultaneously. Reaction time may vary from millisecond scale for ion exchange reactions to days (or months or years) for sorption/desorption reactions to reach equilibrium. Metals present in biosolids-treated soils may be present in different forms and therefore may dissolve at different rates. The data presented in Tables 7-7 to 7-9 and Figures 7-1 to 7-3 may be fitted to a two-site bi-continuum model that includes two dissolution reactions—a fast reaction, which quickly solubilized this component of the metals, followed by a slow reaction of the remaining components that may continue for a long time period. These two reactions may take place simultaneously or in succession. Location and chemical bonding of metals in soil and solution versus metal ratio might be the effects for two pools of metal. The rapid metal releases of the first 15 minutes were fitted into a zero order kinetics model:

$$C_t = k_0 t \qquad \text{for } t = 0 \text{ to } t = 0.25 \qquad \text{[Eq. 7.4-1]}$$

where C_t (mg kg^{-1}) was the metal concentration at time t (hour) and k_0 (mg kg^{-1} hr^{-1}) was the zero order kinetics constant. The remainder of the metal dissolution reaction followed a first order kinetics model:

$$C_t = 0.25\ k_0 + C_s\ [1 - e^{-k_1(t - 0.25)}]; \qquad \text{for } t > 0.25 \qquad \text{[Eq. 7.4-2]}$$

where C_s represented the ultimate metal release due to the first order dissolution reaction (mg kg^{-1}) and k_1 was the first order kinetics constant (hr^{-1}). The kinetics constants depicting the metal dissolution reactions are summarized in Table 7-11. The concentrations of synthetic organic acid mixtures and the amounts of biosolids-borne metals in the soils determined the metals' dissolution and their availability to plants. For the first order slow metal release, the rate constants for a metal (k_1) did not vary significantly.

For the zero order rapid metal release, however, the metal dissolution from the soil increased with the concentration of synthetic organic acid mixtures and the amounts of metals dissolved by the synthetic organic acid mixture of a given concentration increased with amounts of biosolids-borne metals in the soils. When the concentrations of synthetic organic acid mixtures varied from 0.001 to 0.1 M, k_0 for a given soil generally increased by 5-10 times. At the same synthetic organic acid mixture concentration, k_0 of metals varied by 20-100, 15-60, and 10-60 times for control, soil receiving 135 Mg ha^{-1}, and soil receiving 1,080 Mg ha^{-1} of biosolids, respectively. The zero order rate constants are characterized primarily by the metal release in the first 15 minutes. As the percentages of metal released were essentially the same across the control and biosolids treatments (Table 7-10) and the dissolution process was rapid, the difference in rate constants simply reflected the amounts of soluble metals in each soil.

Table 7-11. Kinetics Constants for Metal Dissolution Reactions Calculated According to Equations 7.4-1 and 7.4-2.

Biosolids Treatment[a]	Synthetic Exudate	Zero and First Order Kinetics Constants[b]											
		Cd		Cr		Cu		Ni		Pb		Zn	
		k_0	k_1	k_0	k_1	k_0	k_1	k_0	k_1	k_0	k_1	k_0	k_1
Control	0.001 M	0.01	0.11	0.04	0.09	0.43	0.09	0.24	0.12	n.d.[c]	n.d.	1.2	0.12
	0.01 M	0.03	0.10	0.06	0.09	1.40	0.09	1.56	0.14	n.d.	n.d.	7.8	0.13
	0.1 M	0.08	0.12	0.44	0.08	7.88	0.07	4.28	0.12	0.06	0.09	21.2	0.13
135 Mg ha^{-1}	0.001 M	0.22	0.09	0.28	0.09	3.44	0.09	2.12	0.12	0.84	0.09	15.8	0.11
	0.01 M	0.72	0.12	0.92	0.08	8.84	0.09	6.28	0.12	0.24	0.09	85.6	0.13
	0.1 M	3.16	0.13	3.64	0.09	30.4	0.08	26.8	0.14	1.28	0.08	400	0.13
1,080 Mg ha^{-1}	0.001 M	1.24	0.01	1.36	0.09	9.64	0.08	10.1	0.14	0.52	0.08	100	0.12
	0.01 M	4.64	0.10	3.40	0.08	42.8	0.07	39.2	0.12	1.36	0.08	416	0.12
	0.1 M	6.68	0.07	5.84	0.08	81.2	0.08	64.4	0.12	3.08	0.08	820	0.10

[a]Obtained at Moreno Field Station of the University of California, Riverside, CA from 1976 through 1981 MWRDGC 1975 composted biosolids were applied at rates of 0 (control), 22.5, and 180 Mg ha-1 yr-1, respectively.

[b]Values represent means of two replicates.

[c]Concentration was below detection limit of the instrument for Pb (< 0.001 mg kg^{-1}).

The total metals that were extractable by the synthetic organic acid mixtures were calculated as the sum of metal rapidly released by the zero order reaction (0.25 k_0) and the metal slowly released through the subsequent first order reaction (C_s). The total organic acid extractable metals would be an indication of the amount of metals possibly available to plants. The amounts of metals that were extractable from biosolids-treated soil (1,080 Mg ha^{-1}) by 0.1 M synthetic exudates are presented in Table 7-12.

Table 7-12. Amounts of Metals That Were Extractable from the Biosolids-Treated Romona Soil (1,080 Mg ha^{-1}) by the 0.1 M Synthetic Exudate.

Element	Rapid Release (mg kg^{-1})	Slow Release (mg kg^{-1})	Total Release (mg kg^{-1})
Cd	1.67	0.86	2.53
Cr	1.46	3.74	5.20
Cu	20.3	13.7	34.0
Ni	16.1	8.6	24.7
Pb	0.77	1.73	2.5
Zn	205	80	285

Under this metal release model, plants grown on biosolids-treated soils would be expected to absorb metals from the rapid release pool first. As the metal release followed a zero order reaction, one would expect the plant uptake, therefore tissue concentration, of metals to remain essentially the same until metals in this pool are exhausted. At that time, the plant uptake of metals, and therefore the tissue concentration, would be expected to decrease as metal in the slow release pool becomes available to plants at a much slower rate.

7.4.3 First Order Model

Alternatively, the data presented in Tables 7-7 to 7-9 and illustrated by Figures 7-1 to 7-3 may be treated as time dependent (t) cumulative release (C_t) of metals from biosolids-treated soils for $t = 0 \rightarrow 48$ hours and $C_t = C_0 \rightarrow C_{48}$. The rate of release (or the slope of time period), k_{t-t}, may be expressed as $k_{t-t'} = (C_t - C_{t'})/(t - t')$ or $k_{t-t'} = \Delta C/\Delta t$. As the time intervals for the measurements were not the same throughout the experiment, the present and the previous time steps of the measurements are designated as t and t', respectively. The slope (therefore rate of metal release) progressively decreased over time, fitting the first order reaction model (Table 7-13).

Table 7-13. Rate of Metal Release from the Biosolids-Treated Romona Soil (1,080 Mg ha^{-1}) by 0.1 M Synthetic Exudate.

Element	$k_{0-0.25}$	$k_{0.5-0.25}$	$k_{1-0.5}$	k_{2-1}	k_{4-2}	k_{8-4}	k_{24-8}	k_{48-24}
Cd	6.68	3.40	1.75	0.91	0.48	0.25	0.10	0.05
Cr	5.84	3.56	1.91	1.09	0.63	0.40	0.20	0.11
Cu	81.2	42.4	22.5	11.9	6.30	3.30	1.34	0.70
Ni	64.4	33.6	18.0	9.40	4.88	2.60	0.97	0.50
Pb	3.08	1.64	0.87	0.51	0.27	0.24	0.08	0.05
Zn	820	422	220	113	58.8	30.6	11.5	5.88

The cumulative metal extracted with respect to time was fitted into a single first order kinetics model and can be expressed as:

$$C_t = C_0\,(1 - e^{-k_1 t}) \qquad \text{[Eq. 7.4-3]}$$

where C_t (mg kg^{-1}) was the metal concentration at time t (hour), k_1 (hr^{-1}) was the first order kinetics constant, and C_0 (mg kg^{-1}) represented the ultimate metal release due to the first order dissolution reaction. The concentrations of synthetic organic acid mixtures and the amounts of biosolids-borne metals in the soils determined the metals' dissolution and their availability to plants. For the first order metal release, the rate constants for each metal (k_1) are given in Table 7-14. The amount of metals in the biosolids-treated soils did not appear to affect the outcome.

Table 7-14. First Order Kinetics Constant (k_1) for Metal Dissolution Reaction by Synthetic Organic Acid Mixtures, Calculated According to Equation 7.4-3.

Biosolids Treatment[a]	Synthetic Exudate	First Order Kinetics Constant (k_1)					
		Cd	Cr	Cu	Ni	Pb	Zn
Control	0.001 M	4.71	0.34	4.11	4.87	n.d.[b]	5.00
	0.01 M	3.99	0.30	3.90	3.93	n.d.	5.42
	0.1 M	4.30	0.37	4.49	4.46	0.36	4.92
135 Mg ha^{-1}	0.001 M	4.13	0.38	4.05	4.65	0.38	5.63
	0.01 M	3.88	0.51	4.03	3.95	0.40	5.33
	0.1 M	3.96	0.48	3.81	3.79	0.50	6.06
1,080 Mg ha^{-1}	0.001 M	4.93	0.32	4.30	4.81	0.24	4.85
	0.01 M	5.16	0.40	4.41	4.81	0.32	5.75
	0.1 M	6.25	0.37	4.88	5.53	0.43	6.80

[a]Obtained at Moreno Field Station of the University of California, Riverside, CA. From 1976 through 1981, composted biosolids were applied at dry weight rates of 0 (control), 22.5 and 180 Mg ha^{-1} yr^{-1}, respectively.
[b]Concentration was below detection limits of the AAS for Pb (< 0.001 mg kg^{-1}).

7.5 Summary

The organic acids in the rhizosphere play an important role in solubilizing the metals in biosolids-treated soils and making the metals available for plant absorption. Based on the compositions of organic acids in the rhizosphere of biosolids-treated plant growing media, synthetic organic acid mixtures were formulated to study effects of the compositions, concentrations, and pH of organic acid mixtures on the dissolution of metals in biosolids-treated soils. Findings are summarized as follows:

- Dissolution of metals from biosolids-treated soils was not affected by the compositions of the organic acid mixtures tested. However, the amounts of metals extracted were affected by the concentrations of the organic acid mixtures and their pH.
- In biosolids-treated soils, the metal dissolution rate in the synthetic organic acid mixture followed the first order reaction and, at equilibrium, the amounts of metals dissolved increased in proportion to the volume of organic acids used. In the majority of cases, the metal dissolution increased with decreasing pH for pH < 6.5.
- The synthetic organic acid mixture (synthetic exudate) used in subsequent experiments was based on the average composition of the organic acids recovered from the rhizosphere of the biosolids-treated soils, with a concentration of 0.01 M COO^- in 13.5 mM $Ca(NO_3)_2$, adjusted to pH 4.8.

CHAPTER 8.0

FIELD VALIDATION OF METAL PHYTOAVAILABILITY MODEL FOR BIOSOLIDS-TREATED SOILS

The phytoavailability of a metal in soil customarily is defined by: the amount of metal absorbed by growing plants or by the concentration in the harvested plant tissue. Plants grown on biosolids-treated fields typically absorb much less than 1% of the metals present in the soil. Therefore, estimates of phytoavailability of biosolids-borne metals must account for not only the plant uptake by one single growing season but also the total amounts available over time. Neither the single season plant uptake nor the plant tissue concentration of metals clearly reflects the labile pool of the soil. New approaches are needed to define the phytoavailability of metals in the biosolids-treated soils.

8.1 Model Development

In the rhizosphere, the biosolids-borne metals are present in solid forms and are not readily available to the growing plants. We hypothesize that the pool of phytoavailable metals in biosolids-treated soils is determined by amounts of metals dissolved by organic acids in the rhizosphere. Metal uptake by plants, which determines the plant tissue concentration and phytotoxicity of metals, is determined by the kinetics of metal released into solution by organic acids in the rhizosphere. In this manner, the phytoavailability of biosolids-borne metals may be defined in terms of a capacity factor (organic acids-extractable metals in the rhizosphere), which describes the amount of metals in biosolids-treated soil that will be available over time and a intensity factor (the rate at which metals may be solubilized by organic acids in the rhizosphere), which indicates the rate at which metals are absorbed by plants. Mathematically, the phytoavailability of the biosolids-borne metals may be defined by the following equation:

$$M_t = C \times e^{-k\,t} \quad \text{[Eq. 8.1-1]}$$

where M_t is the amounts of metal removed by crops at year t, C is the capacity factor representing the labile metal pool of the biosolids-treated soils and k is the intensity factor representing the rate metals in the biosolids-treated soils were absorbed by plants. Equation 8.1-1 may also be written in terms of the cumulative metal removal:

$$M = C\,(1 - e^{-kt}) \quad \text{[Eq. 8.1-2]}$$

where *M* is the cumulative metal removal over *t* years of crop harvests. If the capacity and the intensity factors are determined numerically for a biosolids-treated soil, the phytoavailability of the metals in the soil may be defined.

In the laboratory, the organic acid-soluble metals (i.e., *C* in Equations 8.1-1 and 8.1-2) may be determined by extracting the biosolids-treated soils with the synthetic organic acid mixture. When the biosolids-treated soils are extracted with the synthetic organic acid mixture, the amounts of metals entering into the solution are in equilibrium with the amounts of organic acid ligands present, not necessarily representing the total extractable metal. To reach the *C* value in Equations 8.1-1 and 8.1-2, the soil must be successively extracted until the extractable metals in the soil are exhausted. Although the metal dissolution in synthetic organic acid mixture and the rate of metal uptake by plants in the field both follow the first order reaction, the rate constants for metal release in these two cases are not necessarily the same. These two rate constants, however, may be numerically related as the rate of plant uptake is proportional to the rate metals may be solubilized in the rhizosphere. To validate the phytoavailability model for biosolids-borne metals in soils, long-term field data are needed to define the *C* and *k* values and to relate the laboratory-determined capacity and intensity factors to field-observed *C* and *k* values.

8.2 Description of Field Data Used for Model Validation

This section describes the field data that were used in the validation of the model, summarizes experimental results on the metals extractable by the synthetic root exudates (synthetic organic acid mixtures) that determine the capacity factor, fits the field-obtained plant metal uptake and soil metal depletion data to Equation 8.1-1, reconciles the laboratory-determined metal release rate constant with the corresponding metal absorption rate constants obtained from Equation 8.1-1, and confirms the phytoavailability model for the biosolids-treated soils.

8.2.1 Fulton County Land Reclamation Site, Metropolitan Water Reclamation District of Greater Chicago

Metropolitan Water Reclamation District of Greater Chicago (MWRDGC) has maintained a functioning biosolids land application and land reclamation site since the early 1970s. The biosolids application fields are located in Fulton County, IL, approximately 300 km southwest of the Chicago metropolitan area. The fields received annual biosolids applications and, following application, were planted with corn. The biosolids applied were anaerobically digested waste activated sludge withdrawn from mesophilic digesters with a minimum of 15 days retention time and temperature maintained at 35°C. The un-dewatered biosolids were shipped via river barge to the application site, transferred into lagoons, and aged prior to the application. Before 1979, the aged liquid biosolids from the storage lagoons were applied to land by subsurface injection, and afterwards they were surface applied and incorporated into the soil as a cake using a manure spreader as the lagoon gradually dewatered. The biosolids and metal loadings of each field were routinely monitored. At the end of each growing season, the yields and metal concentrations of the harvests were determined. In addition, the soils were sampled to determine the metal concentrations in the soils. This location is designated hereafter as the Fulton County site.

The three fields (Fields 19, 21, and 23) selected for the field validation received, from 1974 to 1984, total mass loadings of 522 to 710 Mg biosolids ha^{-1} dry weight. The cumulative loadings of Cd, Cu, Ni, and Zn are given in Table 8-1. From 1984 through 1997, these fields did not receive any biosolids application and corn was grown and harvested annually. The plants and soil samples collected during 1984-1997 were used in the model validation.

Dr. Thomas C. Granato of Metropolitan Water Reclamation District of Greater Chicago provided the soil and plant tissue samples from which the metal concentrations were determined.

8.2.2 Long-term Biosolids Land Application Experiment, Rosemount Experiment Station, University of Minnesota

The Rosemount Experiment Station, University of Minnesota, was established in 1947. For approximately 30 years prior to application of biosolids in 1976, the experimental field was a part of a corn, small-grain, and forage crop rotation experiment. From 1975 to 1977, biosolids were applied annually on experimental plots at rates of 0, 20, 40, and 60 Mg ha^{-1} yr^{-1}, with four replicates for each treatment. The three biosolids treatments were labeled in ascending order as low, medium and high treatments and the total mass loadings amounted to 60, 120, and 180 Mg ha^{-1}, respectively. The Cd, Cr, Cu, Ni, Pb, and Zn mass loading for the biosolids treatments are given in Table 8-1. With the biosolids applications, plots were maintained in a three-year rotation of corn, alfalfa, and oat. Following the termination of biosolids application, the plots were continuously cultivated with corn annually until 1984. The plant tissue and soil samples collected from 1978 through 1984 were used in the field validation. The description of the experimental set up and the results were summarized in Dowdy et al. (1978). This location is designated hereafter as the Minnesota site.

At the end of each growing season, the total biomass yields were determined and soils were sampled. Corn plant tissue was harvested during the hard dent stage of development and dried at 70°C. Stover and grain tissues were ground separately to pass a 20-mesh sieve. Tissue samples were ashed at 450°C for 24 hours. The ash was equilibrated with 2 M HCl and the supernatant analyzed for Cu, Ni, and Zn by inductively coupled, plasma-source, optical emission spectroscopy (ICP-OES). Cd, Cr, and Pb in the supernatant were determined by flame atomic absorption spectroscopy (AAS).

Dr. Robert H. Dowdy, of the USDA Agricultural Research Station in St. Paul, MN, provided the soil and plant tissue samples that were used in the field validation.

Table 8-1. Cumulative Mass Loading of Biosolids and Trace Elements in Experimental Plots Used for Field Validation – Fulton County and Minnesota Sites.

Site	Treatment	Years Applied	Biosolids (Mg ha^{-1})	Cumulative Loads (kg ha^{-1})					
				Cd	Cr	Cu	Ni	Pb	Zn
Fulton County	Field 19	1974 - 1984	710	177	-	1,142	179	-	2,154
	Field 21	1974 - 1984	681	169	-	1,095	267	-	2,066
	Field 23	1974 - 1984	522	130	-	839	205	-	1,583
Minnesota	Low	1975 – 1978	60	9	48	41	16	54	106
	Medium	1976 – 1977	120	17	91	83	26	97	220
	High	1975 – 1977	180	25	141	127	43	173	343

8.3 Synthetic Organic Acid Mixture Extractable Metals—Determination of Total Available Metals in Biosolids-Treated Soils

The synthetic root exudate, formulated according to the organic acids composition of the rhizosphere (Chapter 7.0), was used to successively extract the metals in the biosolids-treated soil from the Fulton County and Minnesota sites. The soil samples representing the final year of biosolids applications were used to establish the total amount of the organic acid-extractable metals in the biosolids-treated soils at the beginning of the validation period (C in Equation 8.1-1).

For this extraction, aliquots of one-gram soil samples were mixed with 10 mL of synthetic exudate in 50 mL Teflon test tubes. The mixtures were allowed to equilibrate at 25°C using the rotary mixer at approximately 1 rpm for 48 hours. The speed of rotation was maintained constant in all treatments. One mL of chloroform was added to each test tube to prevent microbial decomposition of organic acid mixtures during the course of the equilibration. The extracting solution consisted of 0.01 M organic acids mixture in 13.5 mM $Ca(NO_3)_2$ at pH 4.8. The pH and EC of the system in the reaction period were monitored and an attempt was made to maintain the pH of the system at 4.8. After 48 hours, the extraction approaches equilibrium and the spent extracting solution must be decanted and fresh synthetic solution added to extract additional metals.

In the field, not all of the metals in the biosolids-treated soils are exposed to organic acids present in the rhizosphere. Over time, however, all metals will come into contact with the organic acids produced in the rhizosphere. To determine the total available metals, all of the organic acid-soluble metals of the biosolids-treated soils must be recovered. For this reason, the biosolids-treated soils were extracted with the synthetic exudate for 10 successive times to establish the pattern of metals removal by the extracting solution.

The soil samples of each treatment combination were replicated two times. After the equilibration, the soil suspensions were centrifuged for 20 minutes at 8,000 rpm to separate the solution and solid phases. The solution phase was passed through a 0.45 μm filter paper into 25 mL volumetric flasks. The metal contents of the supernatants were determined using ICP-OES and AAS (Milward and Klukner, 1989). The ICP-OES system was OPTIMA 3000V (Perkin-Elmer, Norwalk, CT, USA) with AS-91 Autosampler and WinLab™ software for the optima family of ICP-OES. The Perkin-Elmer AAnalyst 800 Atomic Absorption Spectrometer (Perkin-Elmer, Bodenseewerk, Germany) with AS-800 Autosampler also was used.

All experiments were replicated. Differences between the experimental groupings were determined by one-way ANOVA ($p < 0.05$) followed by the Student-Newman-Keuls test. Tests were performed with SigmaStat 4.01 Software. To estimate the total available metals and rate constants of the experimental data and k values of the field data, nonlinear regression was used on SigmaStat 4.01 Software.

8.4 Successive Extractions of Metals in Biosolids-Treated Soils

In the field validation, the biosolids-treated soils, that represent the condition of the field at the time the applications were terminated, were successively extracted with the 0.01 M synthetic organic acid mixture. The cumulative metal extracted per unit of soil (i.e., mg metal kg^{-1} soil) may be plotted against the cumulative amounts of organic acids (i.e., moles of COO^-) used in the extraction. Obviously, 10 successive extractions did not remove all of the extractable metals, as illustrated by Figures 8-1 to 8-3. Therefore, the total organic acid extractable metal, which represents the capacity factor of the phytoavailability model, must be projected from the data presented in Figure 8-1 to 8-3.

The release of extractable metals may be hypothesized to follow a first order reaction model in which the rate of the reaction, therefore the amounts of metals extracted in this case, is in proportion to the amounts of extractable metals remaining. With successive extractions, the rate of reaction and the amounts extracted should successively decrease as the remaining extractable metal decreases with each extraction. It is also possible that the release of the extractable metals follows a zero order reaction model in which the rate of metal release remained constant until the available pool is exhausted. Depending on the reaction model selected, the capacity factor determined for the phytoavailability model may be quite different. To determine which reaction model—first order or zero order—is more appropriate to describe the data, the following mathematical procedures were used.

The cumulative metal released by successive extractions may be expressed as Y_0, Y_1, Y_2, Y_3, …Y_n where Y is the cumulative amounts of extracted metals and the subscripts 0, 1, 2, 3, …. n indicate the successive extractions. The corresponding cumulative organic acids used for the extractions may be expressed as X_0, X_1, X_2, X_3, … X_n where X indicates the cumulative amounts of organic acids and the subscripts 0, 1, 2, 3, … n are the successive extractions. The slope of the reaction curve may be expressed successively as:

$$\Delta Y_1/\Delta X_1 = [Y_1 - Y_0]/[X_1 - X_0],$$
$$\Delta Y_2/\Delta X_2 = [Y_2 - Y_1]/[X_2 - X_1],$$
$$\Delta Y_3/\Delta X_3 = [Y_3 - Y_2]/[X_3 - X_2], \ldots$$
$$\Delta Y_n/\Delta X_n = [Y_n - Y_{n-1}]/[X_n - X_{n-1}]$$

For the first order reaction model, the rate of the reaction, as indicated by the slope of the reaction curve, decreases with successive extraction. Therefore, necessary and required criteria for a first order reaction model will be that:

$$\Delta Y_1/\Delta X_1 > \Delta Y_2/\Delta X_2 > \Delta Y_3/\Delta X_3 > \ldots \Delta Y_n/\Delta X_n \rightarrow 0$$

For a zero order reaction model, the rate of reaction remains constant. The necessary and required condition for a zero order reaction model will be that:

$$\Delta Y_1/\Delta X_1 = \Delta Y_2/\Delta X_2 = \Delta Y_3/\Delta X_3 = \ldots \Delta Y_n/\Delta X_n$$

The data on cumulative metal releases vs. cumulative organic acids used to extract soils (representing the year biosolids applications terminated) from the three fields of the Fulton County biosolids land reclamation site and the 16 plots (derived from four treatment levels and four replicates for each treatment level) at the long-term land application experiment at the Minnesota site were evaluated against the criteria above (Tables 8-2 and 8-3). In all of the cases tested, the successive slope ($\Delta Y_n/\Delta X_n$ for $n = 0, 1, 2, \ldots n$) of the cumulative metal release curve series met the necessary and required criteria for a first order reaction. Therefore, the metal removal with respect to successive extractions may be mathematically expressed as

$$M_i = C\,(1 - e^{-k_s i}) \qquad \text{[Eq. 8.4-1]}$$

where M_i depicts the cumulative metals removed after the ith cycle of extractions, C is the ultimate total metal removal, and k_s is the successive extraction rate constant. As additional aliquots of synthetic exudate are introduced with each successive extraction cycle, the parameter i in Equation 8.4-1 indicates the volume of organic acids that has been used in the extraction. Thus, Equation 8.4-1 may be rewritten in terms of the cumulative volume of organic acids used in the extractions as:

$$M_q = a\,(1 - e^{-bq}) \qquad \text{[Eq. 8.4-2]}$$

where M_q represented the cumulative metals removed (mg kg^{-1}) from soil that has been extracted with q cumulative amounts of synthetic organic acid mixture (mol of COO^- kg^{-1}), a represents the ultimate metal removal (mg kg^{-1}), and b is the reaction rate constant (kg mol of $COO^{- -1}$).

Curves generated using these equations, along with the data from Fulton County Field 19, are shown in Figures 8-1 to 8-3. Estimated cumulative metal removal for all three Fulton County fields are given in Table 8-4.

Table 8-2. Successive Slope ($\Delta Y_n/\Delta X_n$) of the Cumulative Metal Release Curve for Biosolids-Treated Soil at Fulton County Biosolids Land Reclamation Site.

Field	Element	$\Delta Y_1/\Delta X_1$	$\Delta Y_2/\Delta X_2$	$\Delta Y_3/\Delta X_3$	$\Delta Y_4/\Delta X_4$	$\Delta Y_5/\Delta X_5$	$\Delta Y_6/\Delta X_6$	$\Delta Y_7/\Delta X_7$	$\Delta Y_8/\Delta X_8$	$\Delta Y_9/\Delta X_9$	$\Delta Y_{10}/\Delta X_{10}$
Field 19	Cd	0.1432	0.1233	0.1151	0.0000	0.1087	0.1069	0.1056	0.1045	0.1037	0.1030
	Cr	0.0268	0.0177	0.0147	0.0131	0.0122	0.0116	0.0112	0.0109	0.0106	0.0104
	Cu	0.2704	0.2019	0.1785	0.1670	0.1602	0.1554	0.1522	0.1498	0.1477	0.1460
	Ni	0.4996	0.3421	0.2912	0.2637	0.2466	0.2357	0.2277	0.2216	0.2169	0.2128
	Pb	0.0292	0.0216	0.0190	0.0176	0.0168	0.0163	0.0160	0.0157	0.0154	0.0152
	Zn	5.9071	3.9195	3.2522	2.9480	2.7345	2.6032	2.5063	2.4475	2.3845	2.3319
Field 21	Cd	0.1328	0.1072	0.0977	0.0932	0.0914	0.0898	0.0886	0.0875	0.0866	0.0860
	Cr	0.0197	0.0129	0.0107	0.0096	0.0090	0.0086	0.0082	0.0079	0.0077	0.0075
	Cu	0.2108	0.1587	0.1409	0.1325	0.1272	0.1237	0.1214	0.1195	0.1180	0.1168
	Ni	0.3930	0.2761	0.2382	0.2182	0.2058	0.1971	0.1814	0.1868	0.1831	0.1803
	Pb	0.0253	0.0167	0.0145	0.0134	0.0127	0.0123	0.0120	0.0118	0.0116	0.0114
	Zn	4.3142	2.8872	2.4410	2.2124	2.0664	1.9653	1.9058	1.8584	1.8166	1.7810
Field 23	Cd	0.1216	0.0951	0.0858	0.0805	0.0784	0.0767	0.0756	0.0744	0.0735	0.0730
	Cr	0.0131	0.0087	0.0071	0.0063	0.0059	0.0055	0.0053	0.0051	0.0050	0.0048
	Cu	0.2016	0.1468	0.1284	0.1195	0.1137	0.1102	0.1077	0.1056	0.1042	0.1028
	Ni	0.3460	0.2424	0.2066	0.1887	0.1779	0.1704	0.1654	0.1614	0.1583	0.1559
	Pb	0.0181	0.0124	0.0104	0.0094	0.0088	0.0084	0.0081	0.0079	0.0077	0.0076
	Zn	3.7168	2.5111	2.1386	1.9469	1.8230	1.7367	1.6877	1.6482	1.6126	1.5796

Table 8-3. Successive Slope ($\Delta Y_n/\Delta X_n$) of the Cumulative Metal Release Curve for Biosolids-Treated Soil at Minnesota Long-Term Biosolids Land Application Site.

Field	Element	$\Delta Y_1/\Delta X_1$	$\Delta Y_2/\Delta X_2$	$\Delta Y_3/\Delta X_3$	$\Delta Y_4/\Delta X_4$	$\Delta Y_5/\Delta X_5$	$\Delta Y_6/\Delta X_6$	$\Delta Y_7/\Delta X_7$	$\Delta Y_8/\Delta X_8$	$\Delta Y_9/\Delta X_9$	$\Delta Y_{10}/\Delta X_{10}$
Low	Cd	0.01400	0.00916	0.00751	0.00666	0.00619	0.00586	0.00562	0.00543	0.00530	0.00519
	Cr	0.00425	0.00333	0.00302	0.00286	0.00276	0.00268	0.00263	0.00259	0.00255	0.00252
	Cu	0.03319	0.02446	0.02140	0.01987	0.01894	0.01832	0.01787	0.01753	0.01726	0.01705
	Ni	0.04947	0.03447	0.02954	0.02708	0.02556	0.02456	0.02384	0.02331	0.02289	0.02255
	Pb	0.01115	0.00735	0.00608	0.00549	0.00513	0.00487	0.00469	0.00454	0.00443	0.00434
	Zn	0.38717	0.27102	0.22935	0.20907	0.19646	0.18805	0.18236	0.17810	0.17453	0.17168
Medium	Cd	0.03748	0.02676	0.02297	0.02092	0.01970	0.01896	0.01841	0.01797	0.01763	0.01736
	Cr	0.00619	0.00522	0.00493	0.00476	0.00464	0.00455	0.00449	0.00444	0.00441	0.00437
	Cu	0.08044	0.05925	0.05202	0.04842	0.04628	0.04483	0.04379	0.04302	0.04241	0.04192
	Ni	0.08361	0.06470	0.05847	0.05524	0.05337	0.05211	0.05119	0.05050	0.04996	0.04953
	Pb	0.01327	0.00907	0.00769	0.00699	0.00656	0.00627	0.00606	0.00591	0.00579	0.00569
	Zn	0.71903	0.53097	0.46460	0.42920	0.40841	0.39491	0.38496	0.37721	0.37119	0.36615
High	Cd	0.04228	0.03022	0.02617	0.02422	0.02300	0.02223	0.02166	0.02122	0.02089	0.02062
	Cr	0.01821	0.01211	0.01007	0.0904	0.0842	0.0800	0.0771	0.0748	0.0731	0.0717
	Cu	0.09553	0.07363	0.06634	0.06242	0.06003	0.05847	0.05735	0.05650	0.05585	0.05532
	Ni	0.11695	0.09243	0.08434	0.08017	0.7774	0.07610	0.7491	0.07402	0.07333	0.07277
	Pb	0.01460	0.01051	0.00917	0.00851	0.00813	0.00786	0.00767	0.00752	0.00740	0.00731
	Zn	0.93363	0.69358	0.90546	0.56582	0.54204	0.52618	0.51580	0.50636	0.49926	0.49258

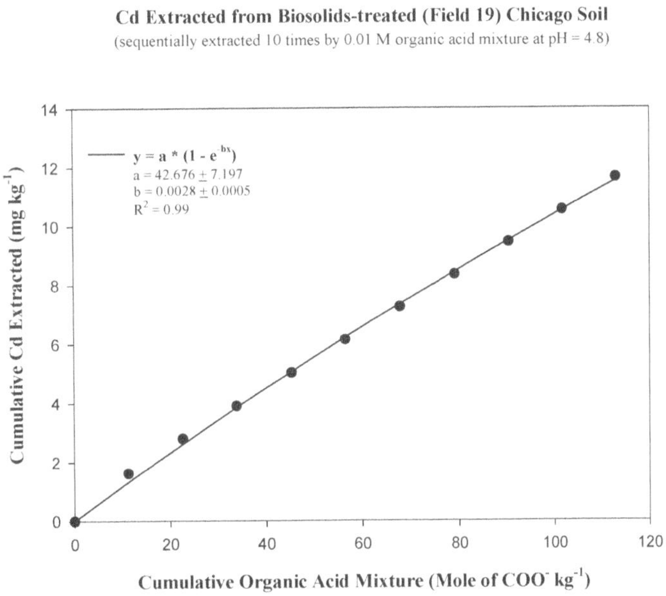

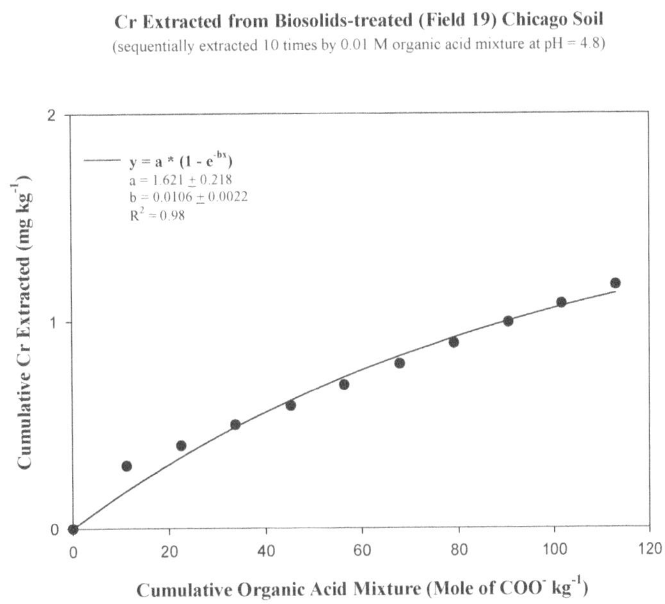

Figure 8-1. Cumulative Cd and Cr Extracted from the Biosolids-Treated Soil of Fulton County Field 19 by Successive Extractions With Synthetic Exudate.

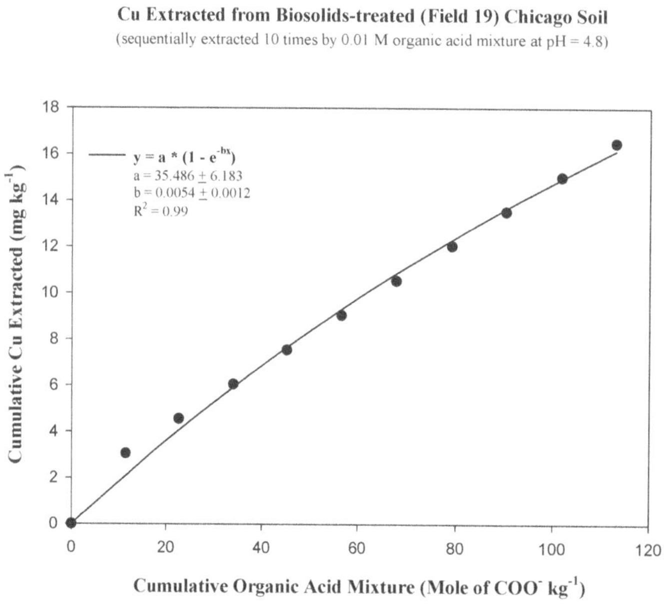

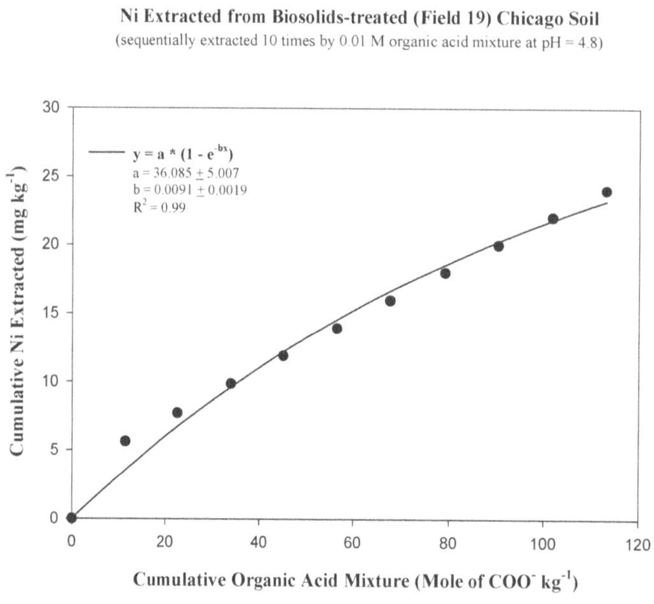

Figure 8-2. Cumulative Cu and Ni Extracted from the Biosolids-Treated Soil of Fulton County Field 19 by Successive Extractions with Synthetic Exudate.

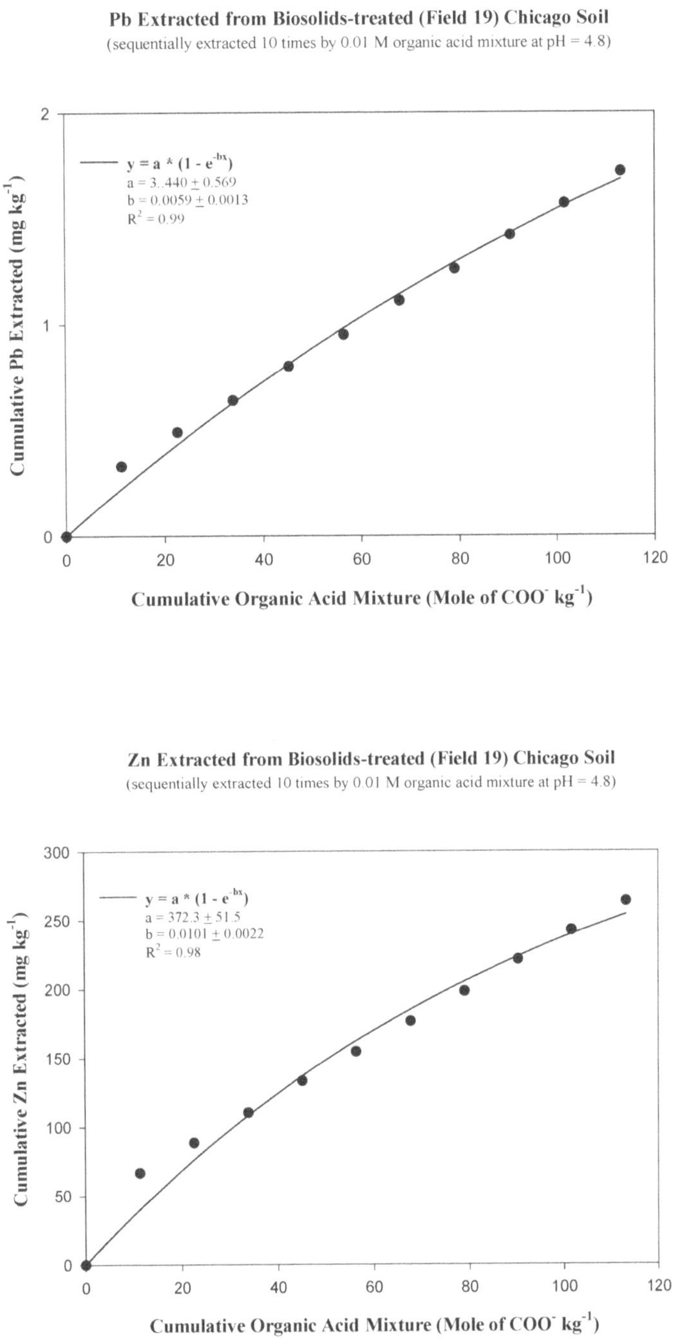

Figure 8-3. Cumulative Pb and Zn Extracted from the Biosolids-Treated Soil of Fulton County Field 19 by Successive Extractions with Synthetic Exudate.

Table 8-4. Estimated Total Available Metals Pool of Fulton County Biosolids-Treated Soils by Successive Extraction Using 0.01 M Synthetic Exudate at pH 4.8, According to Equation 8.4-2.

Field	Element	Total Available ($mg\ kg^{-1}$)[a]	Total Soil Concentration ($mg\ kg^{-1}$)	Extractable Metal (% of total)
Field 19	Cd	42.7 (7.2)	109.3	39
	Cr	1.62 (0.22)	745	0.22
	Cu	35.5 (6.2)	321	11
	Ni	36.1 (5.0)	86.3	42
	Pb	3.44 (0.57)	201	1.7
	Zn	372 (52)	749	50
Field 21	Cd	30.9 (6.1)	81.5	38
	Cr	1.16 (0.15)	588	0.20
	Cu	29.9 (5.4)	231	13
	Ni	33.5 (4.9)	73.3	46
	Pb	2.38 (0.40)	174	1.4
	Zn	299 (43)	616	49
Field 23	Cd	21.8 (4.1)	69.8	31
	Cr	0.70 (0.08)	563	0.12
	Cu	22.9 (3.9)	182	13
	Ni	28. 5 (4.3)	65.8	43
	Pb	1.28 (0.18)	110	1.2
	Zn	275 (40)	516	53

[a]Values represent the mean and standard deviation (in parentheses) of values derived from 3 replicated field plots.

For the biosolids-treated soils from Fulton County, concentrations at the beginning of the validation period (i.e., 1984) were estimated to be 31–39, 0.1–0.2, 11–13, 42–46, 1.2–1.7, and 49–53% of the total Cd, Cr, Cu, Ni, Pb, and Zn, respectively. The metal release rate constants that are represented by the parameter b (in kg per mol of COO^{-}) in Equation 8.4-2 also were estimated (Table 8-5).

Table 8-5. Metal Release Rate Constants for Biosolids-Treated Soils from the Fulton County Site Estimated by Successive Extraction with 0.01 M Synthetic Exudate at pH 4.8, According to Equation 8.4-2.

Field	Element	Release Rate Constant (kg per mol of COO^-)[a]
Field 19	Cd	0.0028 (0.0005)
	Cr	0.0106 (0.0022)
	Cu	0.0054 (0.0012)
	Ni	0.0091 (0.0019)
	Pb	0.0059 (0.0013)
	Zn	0.0101 (0.0022)
Field 21	Cd	0.0033 (0.0007)
	Cr	0.0109 (0.0022)
	Cu	0.0050 (0.0011)
	Ni	0.0079 (0.0016)
	Pb	0.0066 (0.0015)
	Zn	0.0093 (0.0020)
Field 23	Cd	0.0041 (0.0009)
	Cr	0.0122 (0.0024)
	Cu	0.0061 (0.0014)
	Ni	0.0081 (0.0017)
	Pb	0.0092 (0.0020)
	Zn	0.0088 (0.0019)

[a]Values represent the mean and standard deviation (in parentheses) of values derived from 3 replicated field plots.

The metal extraction data for soils of the control and the three biosolids-treated levels (accumulative biosolids loadings of 60, 120, and 180 Mg ha^{-1}) at the Minnesota site also were fitted to Equation 8.4-1 and the results are summarized in Tables 8-6 and 8-7. At the beginning of the validation period (i.e., 1978), 21-24, 1.4-2.0, 7.6-13, 25-39, 2.1-3.3, and 43-47% of the total Cd, Cr, Cu, Ni, Pb, and Zn in the biosolids-treated soils at the Minnesota site were phytoavailable.

Table 8-6. Estimated Total Available Metals of Biosolids-Treated Soils from the Minnesota Site by Successive Extraction with 0.01 M Synthetic Exudate at pH 4.8, According to Equation 8.4-2.

Field	Element	Total Available (mg kg^{-1})[a]	Total Soil Concentration (mg kg^{-1})	Extractable Metal (% of total)
Control	Cd	0.013 (0.007)	0.07	18
	Cr	0.38 (0.05)	28	1.4
	Cu	0.36 (0.04)	4.8	7.4
	Ni	3.59 (0.62)	15	24
	Pb	0.41 (0.07)	16	2.5
	Zn	11.5 (1.5)	27	43
60 Mg ha^{-1} (Low)	Cd	1.07 (0.99)	4.8	22
	Cr	0.68 (0.11)	41	1.6
	Cu	3.73 (0.61)	49	7.6
	Ni	4.26 (0.67)	17	25
	Pb	0.68 (0.09)	33	2.1
	Zn	30.7 (4.6)	71	43
120 Mg ha^{-1} (Medium)	Cd	2.76 (0.49)	13	21
	Cr	1.13 (0.16)	80	1.4
	Cu	9.56 (1.63)	100	9.6
	Ni	14.3 (2.65)	47	30
	Pb	0.99 (0.14)	46	2.1
	Zn	68.1 (12.2)	152	45
180 Mg ha^{-1} (High)	Cd	3.70 (0.72)	16	24
	Cr	1.72 (0.28)	87	2.0
	Cu	14.9 (2.62)	112	13
	Ni	23.4 (4.47)	60	39
	Pb	1.59 (0.27)	48	3.3
	Zn	97.3 (20.6)	207	47

[a]Values represent the mean and standard deviation (in parentheses) of values from 4 replicated field plots.

Table 8-7. Metal Release Rate Constants of Biosolids-Treated Soils from the Minnesota Site Estimated by Successive Extraction with 0.01 M Synthetic Exudate at pH 4.8, According to Equation 8.4-2.

Field	Element	Release Rate Constant (kg per mol of COO^-)[a]
Control	Cd	0.0077 (0.0018)
	Cr	0.0054 (0.0010)
	Cu	0.0110 (0.0022)
	Ni	0.0060 (0.0013)
	Pb	0.0065 (0.0014)
	Zn	0.0103 (0.0021)
60 Mg ha^{-1} (Low)	Cd	0.0116 (0.0025)
	Cr	0.0047 (0.0009)
	Cu	0.0092 (0.0013)
	Ni	0.0077 (0.0017)
	Pb	0.0104 (0.0022)
	Zn	0.0084 (0.0018)
120 Mg ha^{-1} (Medium)	Cd	0.0078 (0.0016)
	Cr	0.0050 (0.0012)
	Cu	0.0059 (0.0013)
	Ni	0.0043 (0.0010)
	Pb	0.0088 (0.0019)
	Zn	0.0065 (0.0014)
180 Mg ha^{-1} (High)	Cd	0.0067 (0.0015)
	Cr	0.0030 (0.0005)
	Cu	0.0047 (0.0010)
	Ni	0.0038 (0.0009)
	Pb	0.0063 (0.0014)
	Zn	0.0056 (0.0013)

[a]Values represent the mean and standard deviation (in parentheses) of 4 replicates.

8.5 Assessment of Field Data Metal Concentrations in the Plant Tissues and in Biosolids-Treated Soils

In the previous section, total phytoavailable metals were estimated for each of the field plots for the initial year of the field validation period. In soils, the chemical forms and phytoavailability of biosolids-borne metals is dependent on the chemical matrix of the receiving soils (Corey et al., 1987b; Sommers et al., 1987). A steady-state condition often is reached in the soils of long-term land application sites such that the chemical forms of the metal in biosolids-treated soils tend to remain stable (Jing and Logan, 1992; Chaney and Ryan, 1993). Therefore, the chemical forms of the metals were assumed not to change during the period the biosolids-treated soils were used for the validation process. From the time at which biosolids application was terminated, the phytoavailable metal pool of the soils should gradually decrease with each crop harvest as metals were absorbed by plants and removed from the soils along with the crop harvests. The removal of the metals from the soils might be tracked by plant metal uptake or by sampling soils following each harvest.

The cumulative metal uptake after t years of plant harvests may be described by the metal phytoavailability model:

$$M = C\,(1 - e^{-kt}) \qquad \text{[Eq. 8.1-2]}$$

where M is the amounts of metal removed by t years of crop harvesting, C is the total available metal pool of the biosolids-treated soil (represented by total synthetic root exudates-extractable metals, a) in Tables 8-5 to 8-7, and k is the plant uptake rate constant of the estimated available metal pool. To distinguish it from the metal phytoavailability model, the equation used to fit the field validation data are described as follows:

$$M_f = C\,(1 - e^{-kt}) \qquad \text{[Eq. 8.5-1]}$$

where M_f is the field-derived cumulative metal removal (mg kg^{-1}), k is the plant uptake rate constant (yr^{-1}), and t is years following estimation of the original available pool.

Metal uptake by plants was determined for Fields 19, 21, and 23 at the Fulton County site (Figures 8-4 and 8-5). From 1984, when biosolids applications were terminated, through 1997, the end of the field validation data set, the corn harvested each year cumulatively removed 1.85-2.25, 4.3-4.5, 0.7-1.0, and 45-50 kg ha^{-1} of Cd, Cu, Ni, and Zn respectively. Based on the depth of sampling and bulk density of soils, these removals may be converted to a decrease in the metal concentrations in the soils and fitted to Equation 8.5-1 to obtain the k for each field (Table 8-8).

The duration for which biosolids-borne metals will be available to plants may be estimated by the rate constants k_l, in Table 8-8. The half-lives of the estimated bioavailable metal pool ($t_{1/2}$), were estimated as

$$t_{1/2} = 0.693/k \qquad \text{[Eq. 8.5-2]}$$

Based on the calculated $t_{1/2}$ values (Table 8-8), an estimated 329-448, 210-273, 280-378, 168-252, 154-182, and 112-161 years of continuing cultivation would be required to remove >99% of the phytoavailable Cd, Cr, Cu, Ni, Pb, and Zn, respectively, by plant harvests.

Table 8-8. Estimated Rate Constants (k) and Corresponding Metals Depletion Half-lives for Biosolids-Treated Soils at the Fulton County Site, According to Equations 8.5-1 and 8.5-2.

Field Location	Element	k ($year^{-1}$)	Half-life, $t_{1/2}$ (year)
Field 19	Cd	0.0108	64
	Cr	0.0178	39
	Cu	0.0129	54
	Ni	0.0191	36
	Pb	0.0271	26
	Zn	0.0298	23
Field 21	Cd	0.0116	60
	Cr	0.0194	36
	Cu	0.0137	51
	Ni	0.0209	33
	Pb	0.0304	23
	Zn	0.0341	20
Field 23	Cd	0.0146	47
	Cr	0.0231	30
	Cu	0.0174	40
	Ni	0.0283	24
	Pb	0.0317	22
	Zn	0.0421	16

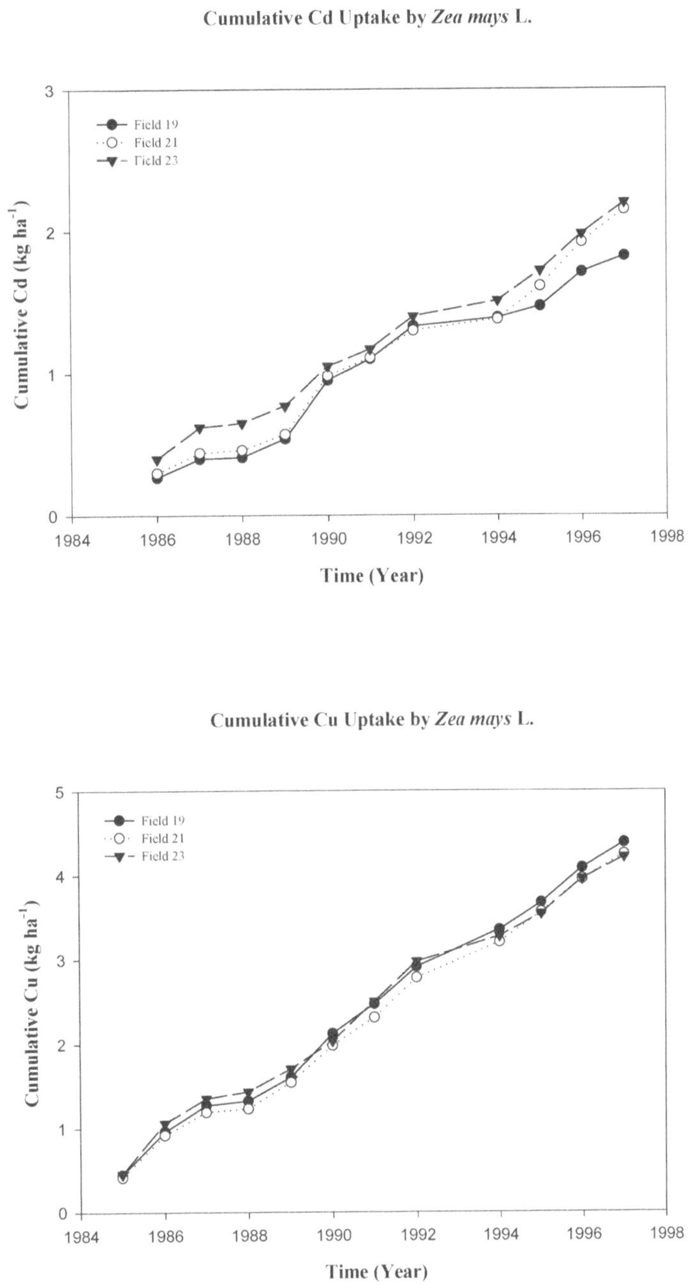

Figure 8-4. Cumulative Cd and Cu Uptake by Corn Harvested from Biosolids-Treated Fields at the Fulton County Site, Following the Termination of Biosolids Applications in 1984.

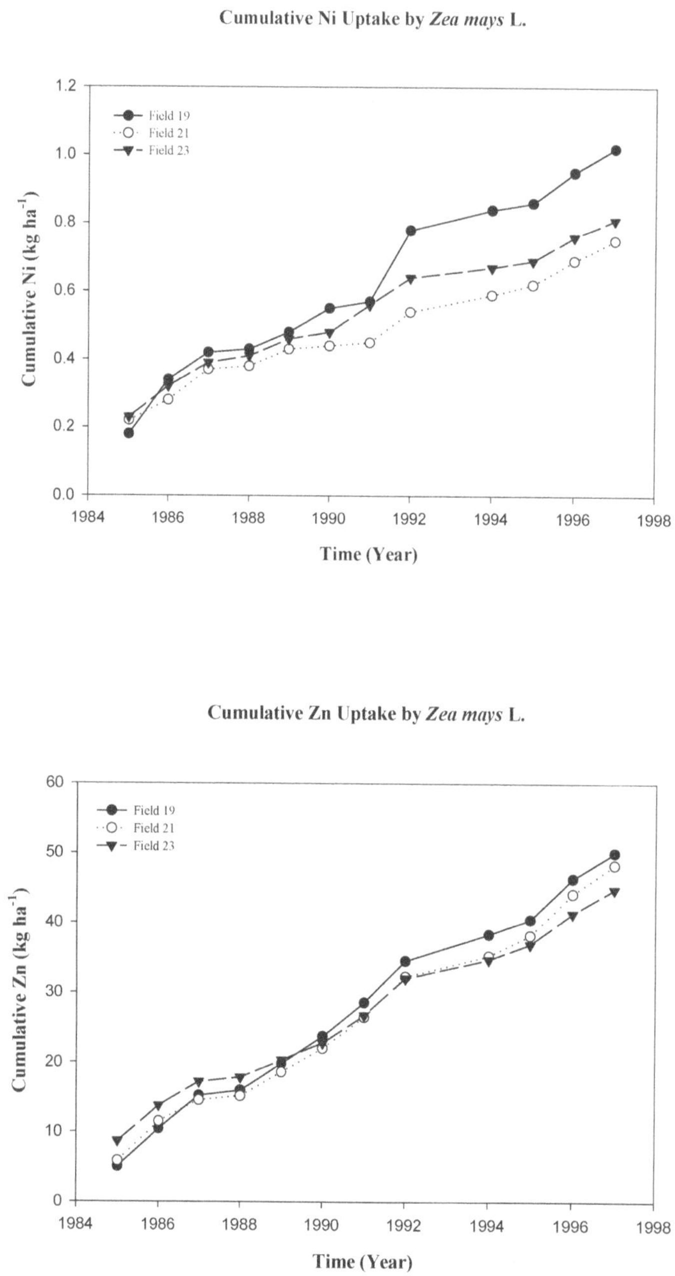

Figure 8-5. Cumulative Ni and Zn Uptake by Corn Harvested from Biosolids-Treated Fields at the Fulton County Site, Following the Termination of Biosolids Application in 1984.

The plant uptake of metals for long-term land application experiments at the Minnesota site (Figures 8-6 to 8-11) also were treated in the same manner. The total synthetic exudates-extractable metals in the biosolids-treated soils from 1978 (the year biosolids treatments terminated) were determined according to the successive extractions and the fitting of the data to the first order reaction model (Table 8-6). Again, this procedure established the phytoavailable metal pools of the biosolids at the beginning of the field validation period. From this point on, the available metals pools steadily decreased as metals were absorbed by plants and removed with crop harvests. During the validation period, cumulative metal removal over time may be described by Equation 8.3. By converting the plant metal uptake data in Figure 8-6 to 8-11 from the units of mass per unit production area (kg ha^{-1}) to mass per unit soil (mg kg^{-1}), the field-based metal removal rate constant (k) and metal depletion half-lives may be estimated (Table 8-9). For the Minnesota site, the total metal inputs from biosolids land application were considerably lower that those at the Fulton County site, the depletion rate for phytoavailable metals were similar and the estimated time required for >99% of the phytoavailable metal to be removed by the plant harvests are 133-217, 175-469, 133-469, 182-350, 98-161, and 98-140 for Cd, Cr, Cu, Ni, Pb, and Zn, respectively.

Table 8-9. Estimated Metal Removal Rate Constants (k) and Corresponding Metals Depletion Half-lives for Biosolids-treated Soils at the Minnesota Site, According to Equations 8.5-1 and 8.5-2.

Treatment	Element	k ($year^{-1}$)	Half-life (year)
Control	Cd	0.0272	25
	Cr	0.0278	25
	Cu	0.0331	21
	Ni	0.0290	24
	Pb	0.0333	21
	Zn	0.0494	14
60 Mg ha^{-1}	Cd	0.0363	19
	Cr	0.0278	25
	Cu	0.0370	19
	Ni	0.0270	26
	Pb	0.0481	14
	Zn	0.0492	14
120 Mg ha^{-1}	Cd	0.0226	31
	Cr	0.0112	62
	Cu	0.0159	44
	Ni	0.0170	41
	Pb	0.0413	17
	Zn	0.0328	21
180 Mg ha^{-1}	Cd	0.0223	31
	Cr	0.0104	67
	Cu	0.0103	67
	Ni	0.0140	50
	Pb	0.0306	23
	Zn	0.0344	20

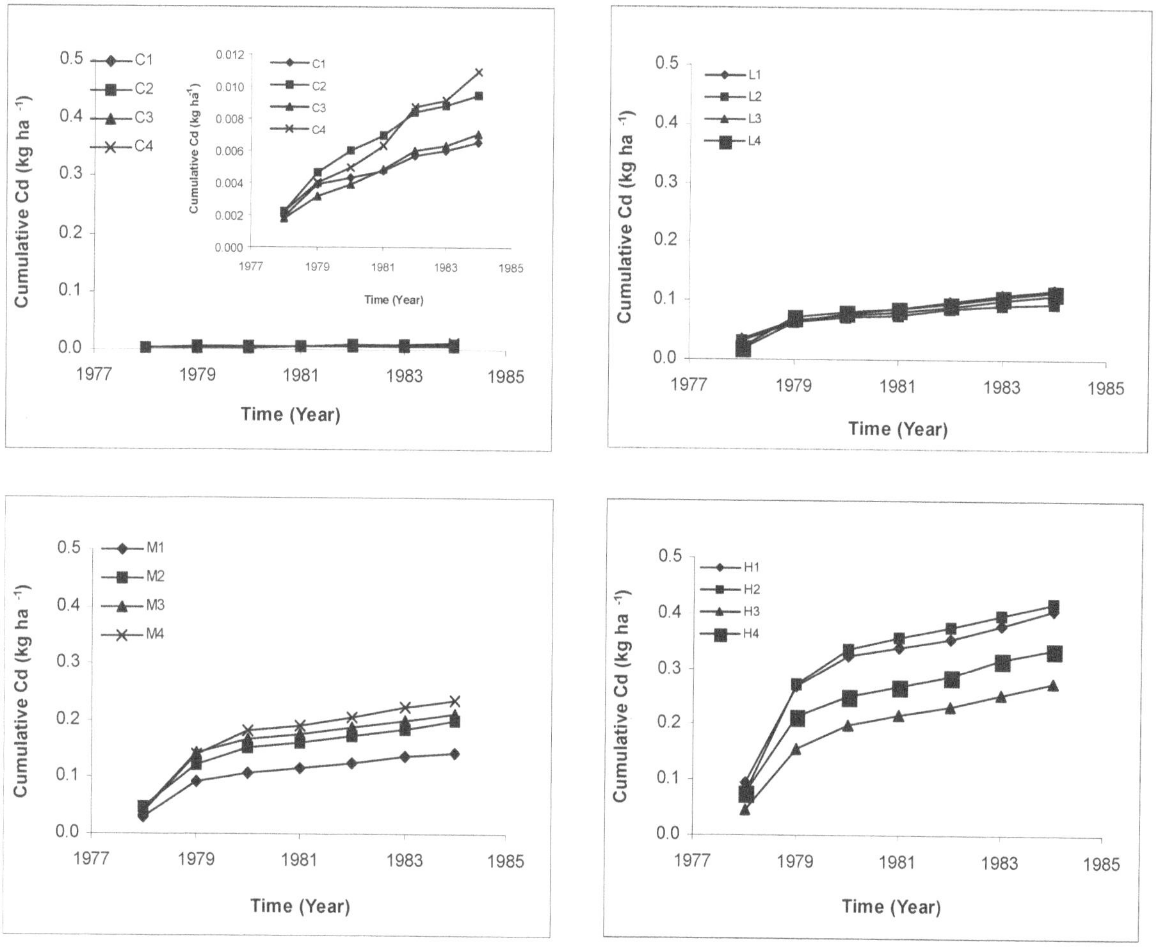

Figure 8-6. Cumulative Cd Uptake by Corn Harvested from Biosolids-Treated Plots at the Minnesota Site, Following the Termination of Biosolids Application in 1978. [Figure legends C1 to C4, L1 to L4, M1 to M4, and H1 to H4 denote respectively the 4 replicates of Control (0 Mg ha^{-1}), Low (60 Mg ha^{-1}), Medium (90 Mg ha^{-1}), and High (120 Mg ha^{-1}) Biosolids Treatments of the Soils.]

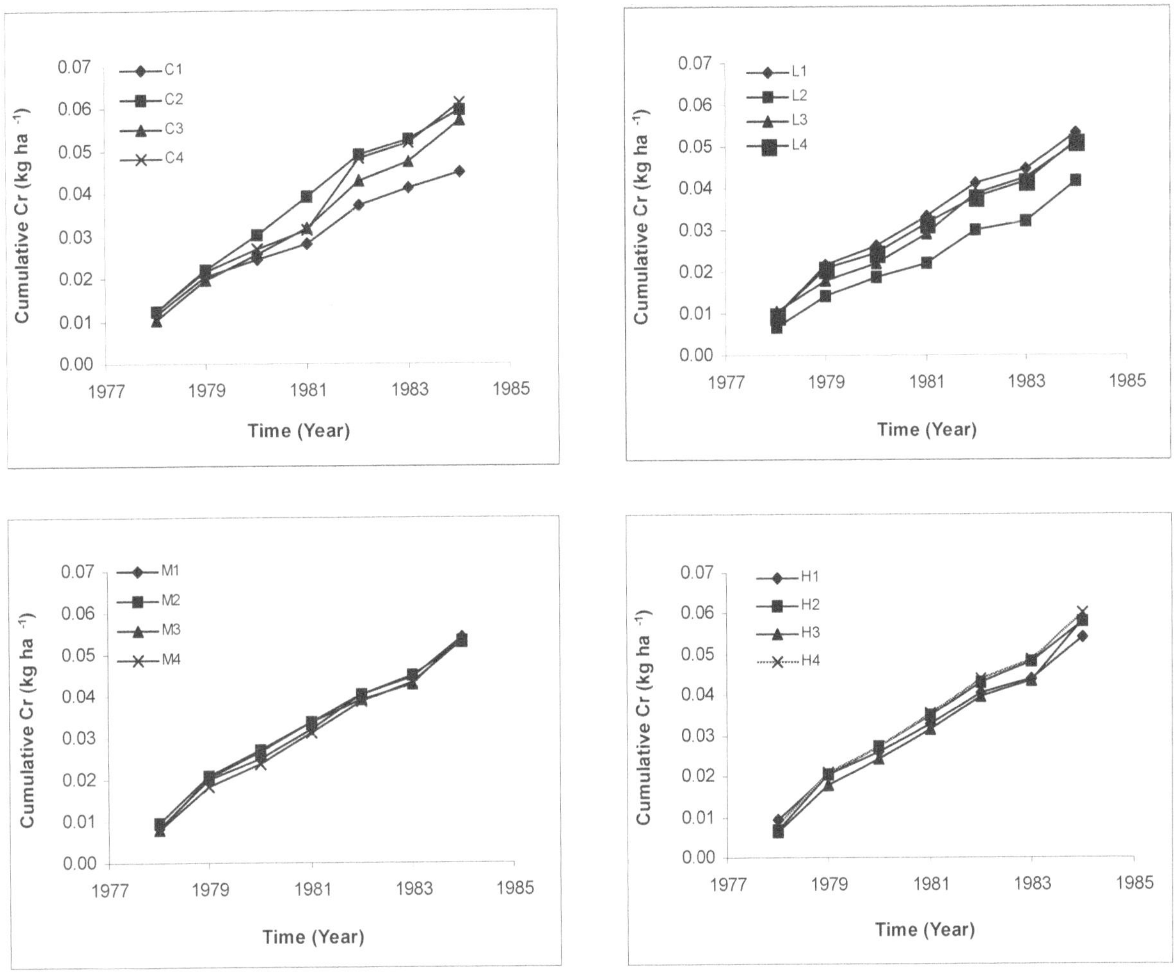

Figure 8-7. Cumulative Cr Uptake by Corn Harvested from Biosolids-Treated Plots at the Minnesota Site, Following the Termination of Biosolids Application in 1978. [Figure legends C1 to C4, L1 to L4, M1 to M4, and H1 to H4 denote respectively the 4 replicates of Control (0 Mg ha^{-1}), Low (60 Mg ha^{-1}), Medium (90 Mg ha^{-1}), and High (120 Mg ha^{-1}) Biosolids Treatments of the Soils.]

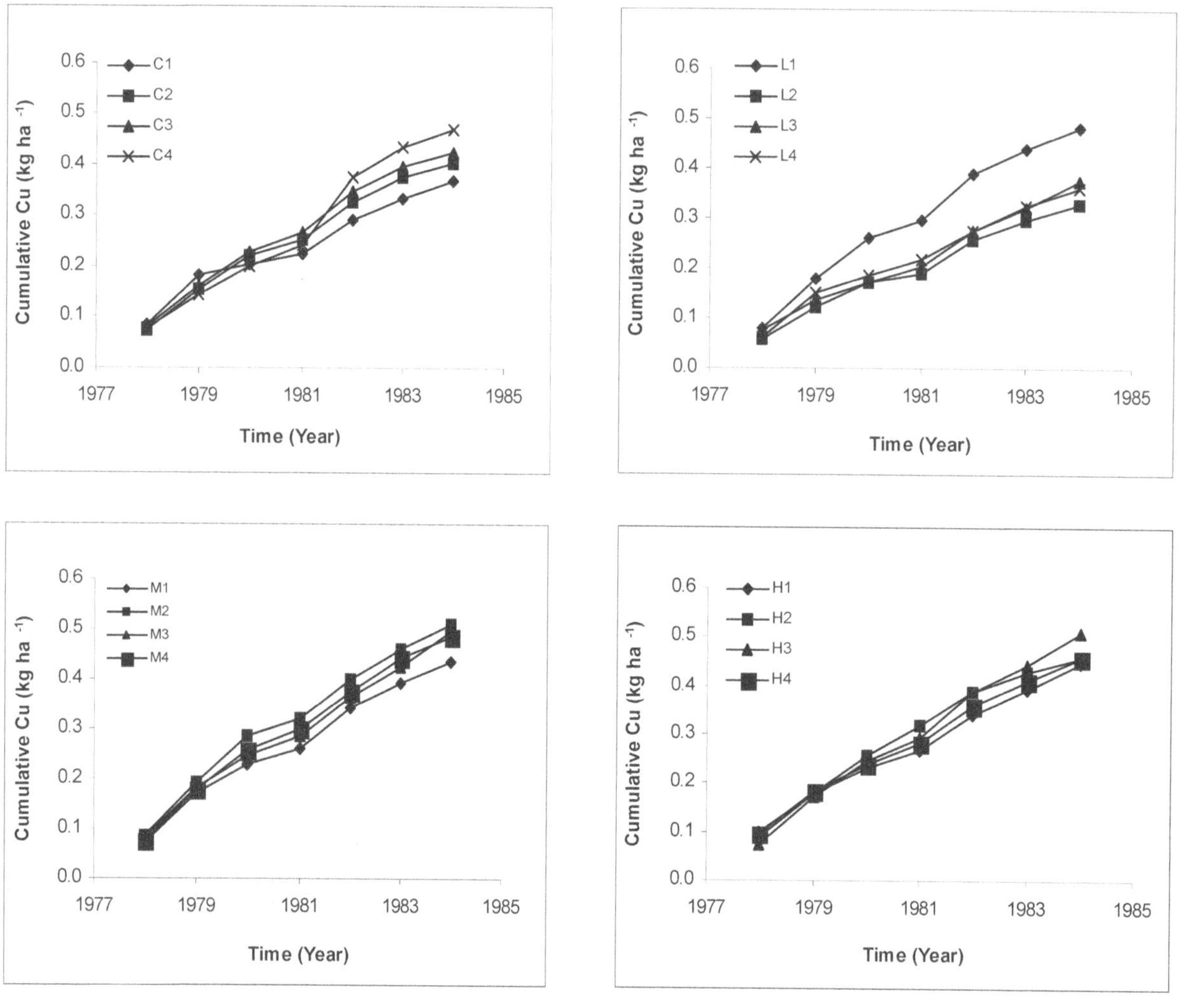

Figure 8-8. Cumulative Cu Uptake by Corn Harvested from Biosolids-Treated Plots at the Minnesota Site, Following the Termination of Biosolids Application in 1978. [Figure legends C1 to C4, L1 to L4, M1 to M4, and H1 to H4 denote respectively the 4 replicates of Control (0 Mg ha^{-1}), Low (60 Mg ha^{-1}), Medium (90 Mg ha^{-1}), and High (120 Mg ha^{-1}) Biosolids Treatments of the Soils.]

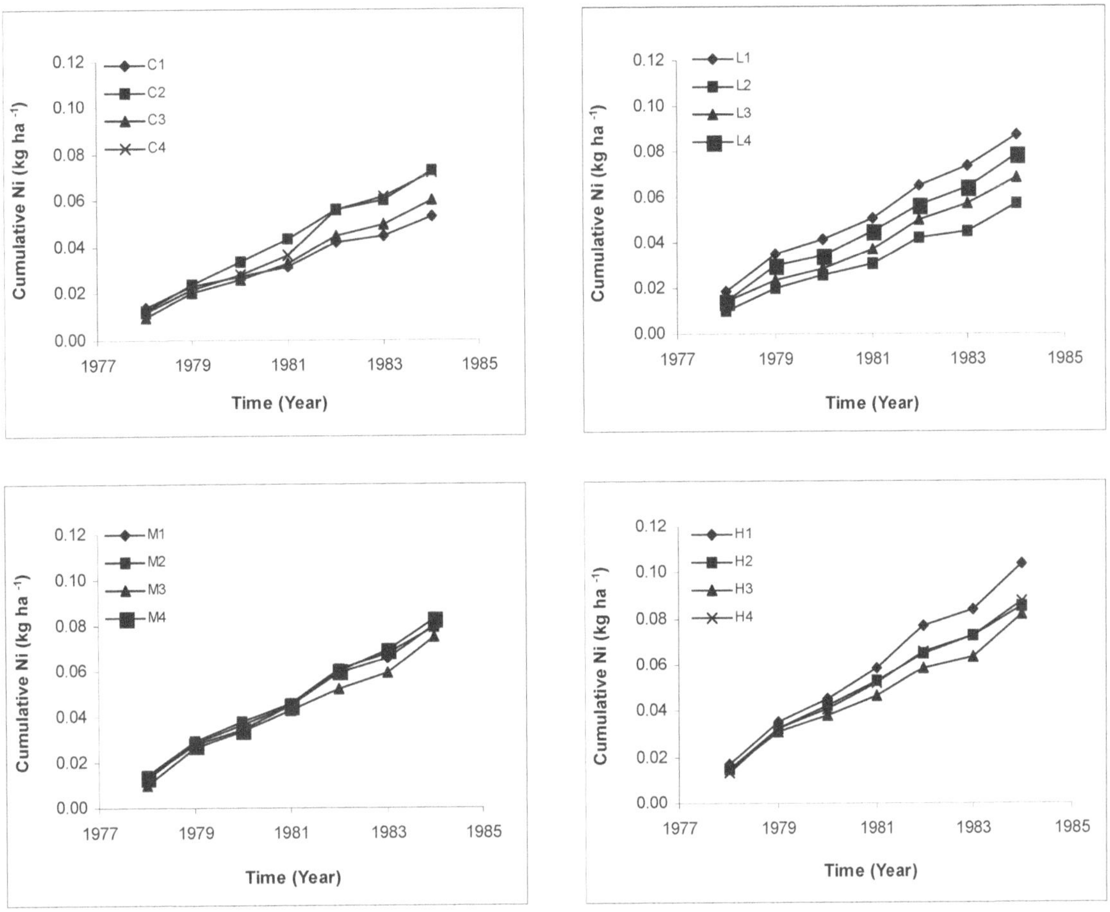

Figure 8-9. Cumulative Ni Uptake by Corn Harvested from Biosolids-Treated Plots at the Minnesota Site, Following the Termination of Biosolids Application in 1978. [Figure legends C1 to C4, L1 to L4, M1 to M4, and H1 to H4 denote respectively the 4 replicates of Control (0 Mg ha^{-1}), Low (60 Mg ha^{-1}), Medium (90 Mg ha^{-1}), and High (120 Mg ha^{-1}) Biosolids Treatments of the Soils.]

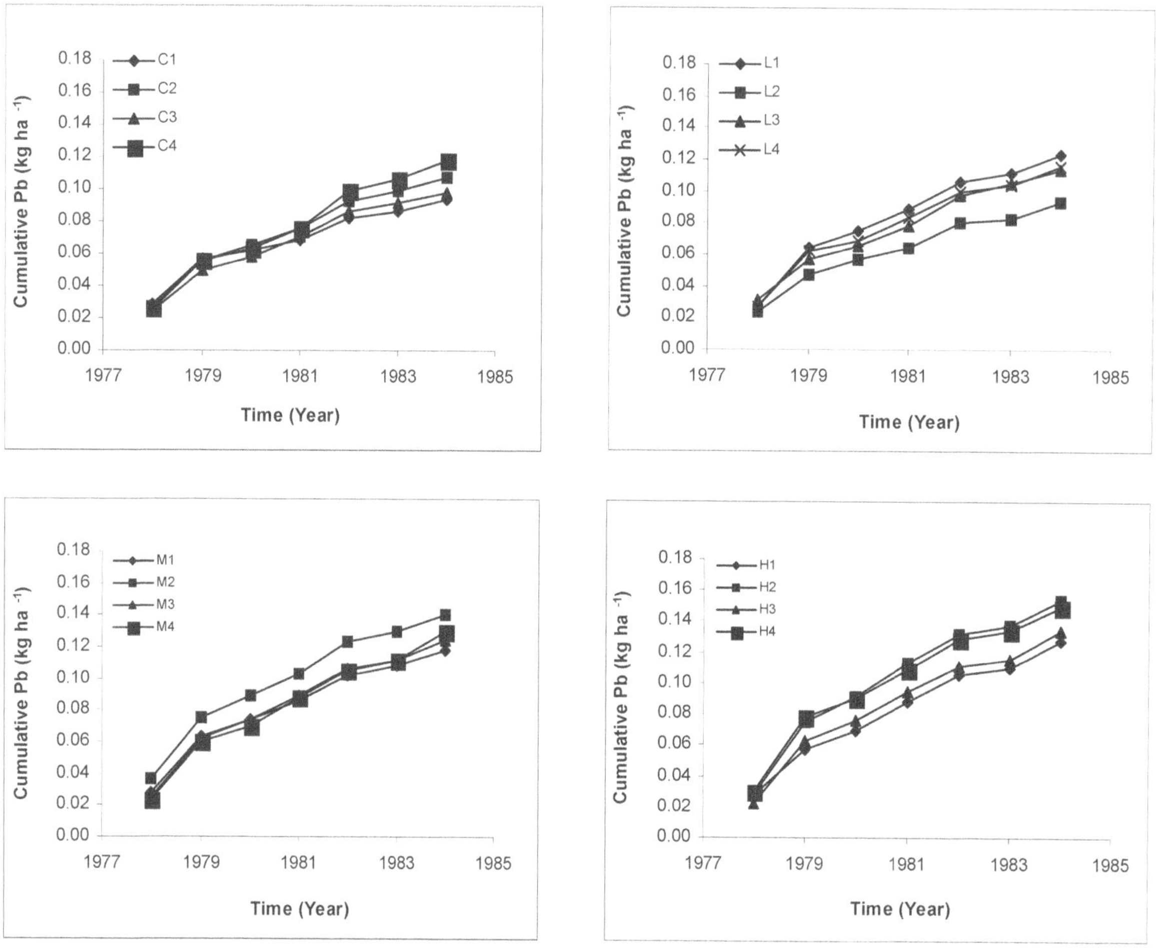

Figure 8-10. Cumulative Pb Uptake by Corn Harvested from Biosolids-Treated Plots at the Minnesota Site, Following the Termination of Biosolids Application in 1978. [Figure legends C1 to C4, L1 to L4, M1 to M4, and H1 to H4 denote respectively the 4 replicates of Control (0 Mg ha^{-1}), Low (60 Mg ha^{-1}), Medium (90 Mg ha^{-1}), and High (120 Mg ha^{-1}) Biosolids Treatments of the Soils.]

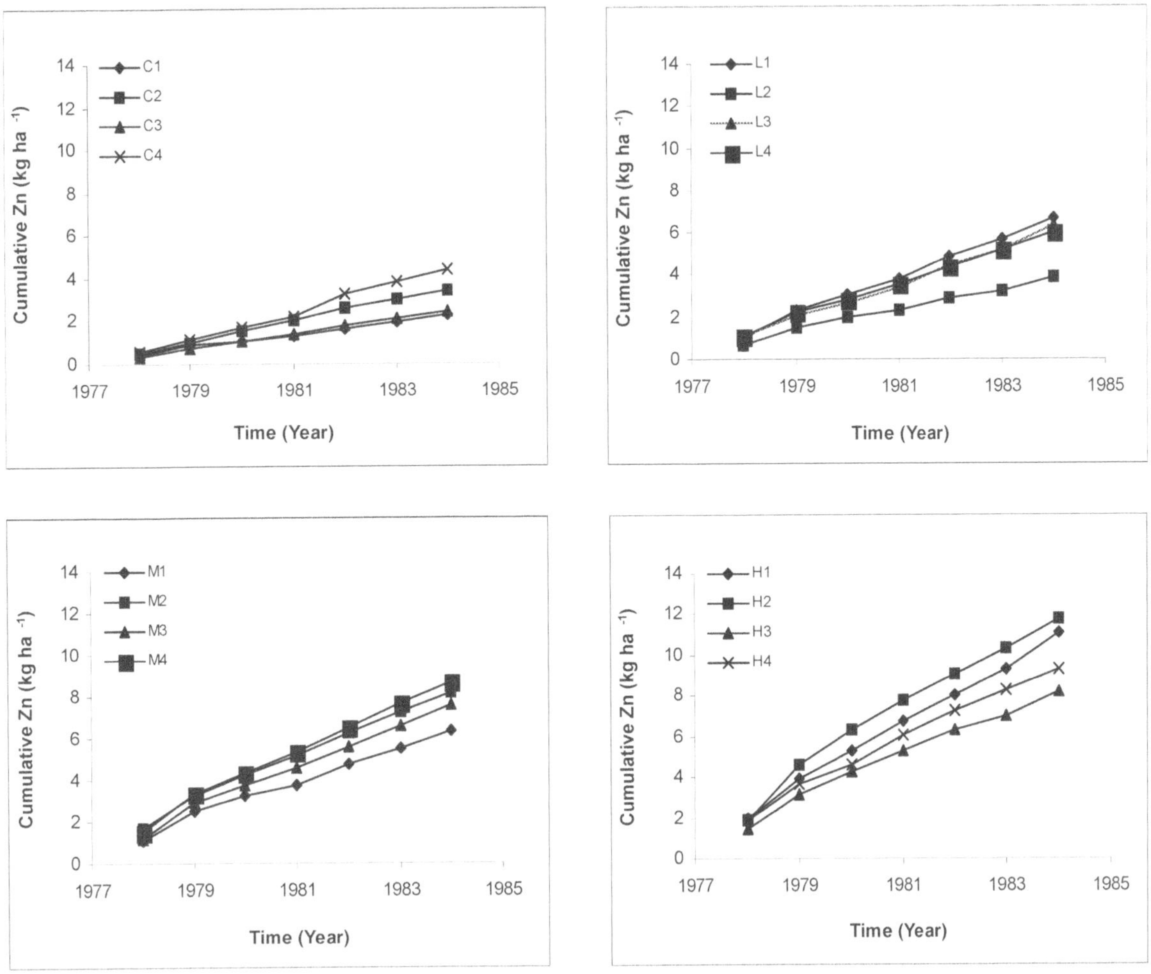

Figure 8-11. Cumulative Zn Uptake by Corn Harvested from Biosolids-Treated Plots at the Minnesota Site, Following the Termination of Biosolids Application in 1978. [Figure legends C1 to C4, L1 to L4, M1 to M4, and H1 to H4 denote respectively the 4 replicates of Control (0 Mg ha^{-1}), Low (60 Mg ha^{-1}), Medium (90 Mg ha^{-1}), and High (120 Mg ha^{-1}) Biosolids Treatments of the Soils.]

8.6 Dissolution Rate of Metals in Biosolids-Treated Field Soils

In the field validation process, the objectives were to determine the total phytoavailable metal pool (C) of the biosolids-treated soils, and the corresponding plant uptake rate constant (k) values, independently. In this manner, the parameters in the metal phytoavailability model [$M = C\,(1 - e^{-kt})$], may be defined. As described in Section 8.3, total phytoavailable metals of the biosolids-treated soils were obtained by successively extracting soils with a synthetic exudate and fitting the data to a first order reaction model. At the same time, the synthetic exudate-based metal dissolution rate constants also were determined. These rate constants (Tables 8-5 and 8-7) had units of kg per mole of COO^- and represented the rate at which the metals were dissolving from soils, and thus becoming available, with successive exposures to organic acids in the synthetic root exudates. In Section 8.4, the field-based plant metal uptake data and the laboratory-determined total phytoavailable metal values were used to obtain the field-based plant metal uptake rate constant (Table 8-8 and 8-9), which has units of (yr^{-1}). These two rate constants represented the release of metal from biosolids-treated soils by organic acids. They can be mathematically reconciled, thereby allowing estimation of the field-based metal uptake rate constants (parameter k in Tables 8-8 and 8-9) from the laboratory-determined synthetic exudate-soluble metal release rate constant (Tables 8-5 and 8-7).

In the laboratory successive extractions, metals in the biosolids-treated soils were exposed to a large volume of organic acids in a short period of time and metals were released rapidly. In the field, the metals in the biosolids-treated soils were exposed to small amounts of slowly released organic acids in the rhizosphere over time of the plant growth. While not all of the metal containing particles in the rhizosphere are exposed to the organic acids at the same time, over time all of the metal containing particles will come into contact with the organic acids. Therefore, the amounts of metal released in both cases should be proportional to amounts of organic acids reacting with metals in the soils. As the volume of organic acids increases, both the amount of metal released through successive extractions (in the laboratory situation) and metal uptake by the plant (in the field situation) increase.

Metal uptake by plants is described by

$$M_t = C \times e^{-kt} \qquad \text{[Eq. 8.1-1]}$$

where M_t is the metal remaining in soil after time t of the plant growth. Organic acids produced in the rhizosphere are proportional to duration of plant growth. In the meantime, the metal released by successive extractions, described by

$$M_q = a\,(1 - e^{-bq}) \qquad \text{[Eq. 8.4-2]}$$

may be rewritten as:

$$M_Q = a\,(e^{-bq}) \qquad \text{[Eq. 8.6-1]}$$

where M_Q represents the metals remaining in the soil after successive extractions by q volumes of organic acids.

Let A be the effective annual organic acids production (mole of COO^- kg^{-1} yr^{-1}) of the rhizosphere. For plants to produce an equivalent of q volume of organic acids, it will take:

$$q = A\,t \qquad \text{[Eq. 8.6-2]}$$

where q is the volume of organic acids (mole COO^- kg $^{-1}$) and t (yr) is the time of plant growth. Substituting q from Equation 8.6-2, Equation 8.6-1 may be converted into the time-equivalent metal release equation:

$$M_{Qt} = a\,(e^{-bAt}) \qquad \text{[Eq. 8.6-3]}$$

where M_{Qt} is the time equivalent metal remaining in soil after successive extraction with q volume of organic acids. The remaining parameters have been previously defined. Following this conversion, Equations 8.1-1 and 8.6-3 represent the same metal release process and $M_t = M_{Qt}$. Therefore,

$$C\,(e^{-kt}) = a\,(e^{-bAt}) \qquad \text{[Eq. 8.6-4]}$$

As C is defined to equal a,

$$kt = bAt \qquad \text{[Eq. 8.6-5]}$$

and

$$k = bA \qquad \text{[Eq. 8.6-6]}$$

In this manner, the field-based metal uptake coefficient (k) is mathematically related to the successive extraction metal release rate constant (b) through the effective field organic acid production (A). For a given biosolids-treated field, the parameter, A, may be defined by k and b:

$$A = k/b \qquad \text{[Eq. 8.6-7]}$$

The effective annual organic acids production of the rhizosphere (A) for experimental fields at the Fulton County and Minnesota sites were calculated (Tables 8-10 and 8-11).

Table 8-10. Estimated Effective Annual Organic Acids Production of the Rhizosphere (*A*) for Biosolids-Treated Soils at the Fulton County Site.

Field Location	Effective Organic Acids Production[a] (mol kg^{-1} year^{-1})					
	Cd	Cr	Cu	Ni	Pb	Zn
Field 19	3.9 Aa	1.7 Ab	2.4 Aa	2.1 Aa	4.6 Aa	3.0 Aa
Field 21	3.5 Aa	1.8 Ab	2.7 Aa	2.7 Aa	4.6 Aa	3.7 Aa
Field 23	3.6 Aa	1.9 Ab	2.9 Aa	3.5 Aa	3.5 Aa	4.8 Aa

[a]Values summarized in this table represent means of four replicates. The differences among the fields were tested by one-way ANOVA. In each column, values followed by the same upper case letter were not significantly different at $p < 0.05$. The differences of the A values between metals were tested by Student-Newman-Keuls test. In each row, values followed by different lower case letter were significantly different at $p < 0.05$.

Table 8-11. Estimated Effective Annual Organic Acids Production of the Rhizosphere (*A*) for Biosolids-Treated Soils at the Minnesota Site.

Soil	Effective Organic Acids Production Factor[a] (mol kg^{-1} $year^{-1}$)					
	Cd	Cr	Cu	Ni	Pb	Zn
Control	3.5 Aa	5.2 Aa	3.0 Aa	4.8 Aa	5.1 Aa	4.8 Aa
60 Mg ha^{-1}	3.1 Aa	5.9 Aa	4.0 Aa	3.5 Aa	4.6 Aa	5.9 Aa
120 Mg ha^{-1}	2.9 Aa	2.2 Ba	2.7 Aa	4.0 Aa	4.7 Aa	5.1 Aa
180 Mg ha^{-1}	3.3 Aa	3.5 Aa	2.2 Aa	3.7 Aa	4.9 Aa	6.1 Aa

[a]Values represent means of four replicates. The differences among the fields were tested by one-way ANOVA. In each column, values followed by the same upper case letter were not significantly different at $p < 0.05$. The differences of the A values between metals were tested by Student-Newman-Keuls test. In each row, values followed by different lower case letter were significantly different at $p < 0.05$.

The effective organic acid production factors (*A*) were determined independently for each treatment combination. For a given location, the range of the *A* values for an element were relatively narrow. Even between the elements, the average *A* values among the elements were similar. The data, summarized in Tables 8-10 and 8-11, were tested statistically for differences between treatments and metal elements (at $p < 0.05$) using the Student-Newman-Keuls test. Only the *A* values for Cr at the Fulton County site were significantly different from other elements at the same field. As the results indicated that A values due to biosolids treatments were not significantly different at $p < 0.05$, the data for all levels of biosolids treatments were pooled to test the difference between the elements. When the data were pooled, the results indicated that differences due to elements were no longer significantly difference at $p < 0.05$. The data for all elements were then pooled to test for differences between soils. The results indicated that the effective organic acid production factors (*A*) were not significantly different between the two locations.

The statistical inferences indicated that the effective organic acid production factor for a plant species (corn in this case) did not vary significantly due to biosolids treatments, metal elements, or locations. The entire population of *A* values (based on a total of168 observations) were pooled to determine the probabilistic characteristics of the population (Figures 8-12 and 8-13).

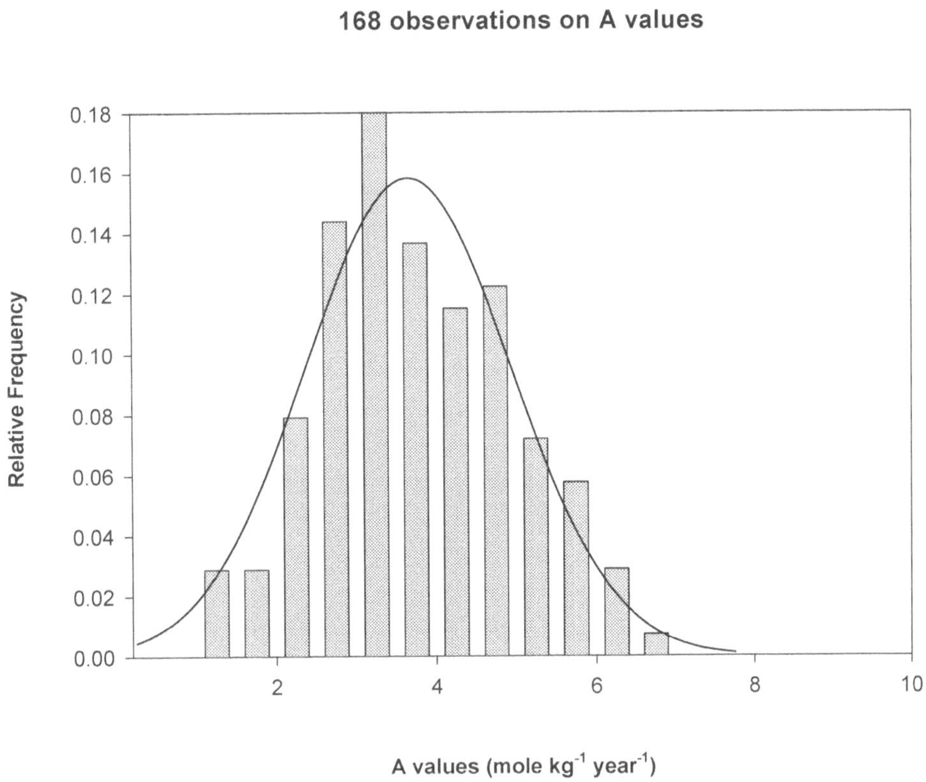

Figure 8-12. Frequency Distribution of the Effective Annual Organic Acid Production of the Rhizosphere of Corn (Mean = 3.80, Standard Deviation = 1.20, and Variance = 1.45).

Cumulative Probability on A values (168 observations)

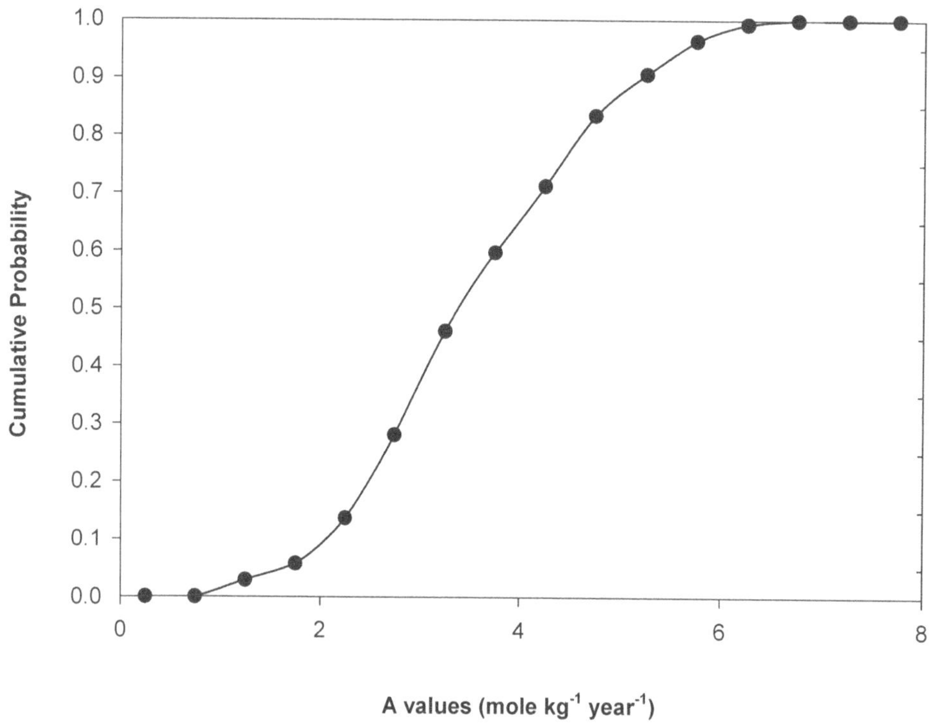

Figure 8-13. Cumulative Probability for the Effective Annual Organic Acid Production (*A*) for Corn.

The population of 168 observations is normally distributed (Figure 8-12) and the probability density function of its distribution can be expressed as:

$$f(x) = \frac{1}{\sigma\sqrt{2\pi}\, e^{\frac{-(x-\mu)^2}{2\sigma^2}}} \qquad \text{[Eq. 8.6-8]}$$

where σ is the standard deviation, μ is the mean, and σ^2 is the variance. The probability density function of the effective organic acid factor (A) is presented in Figure 8-13.

Based on the probability distribution (Figure 8-13), an A value may be selected. In this manner, the metal phytoavailability estimates derived from the A value selected will be linked to a probability that the actual metal phytoavailability will be equal to or below the estimate.

Table 8-12. Probability Density Function of the Effective Organic Acid Factors (A) Estimated from Each Metal of the Fulton County and Minnesota Sites from One-stage Model.

A Value (mol kg^{-1} year^{-1})	1	2	3	4	5	6	7	8
Probability (%)	0.99	6.81	25.5	56.8	84.1	96.6	99.6	100

8.7 Summary

The phytoavailability of heavy metals in the biosolids-treated soils is described by the following equation

$$M_t = C\,(1 - e^{-kt}) \qquad \text{[Eq. 8.1-2]}$$

where M_t (mg kg $^{-1}$) is the cumulative metal removal from the biosolids-treated soils by growing and harvesting plants for t years, C is the total phytoavailable metal pool (mg kg $^{-1}$), and k is the metal release coefficient (yr $^{-1}$) of the soils. If the C and k of a biosolids-treated soil are defined, the phytoavailability of the metals in the biosolids-soil matrix may be evaluated.

For biosolids-treated soils, the C value may be determined by successively extracting the soil with a synthetic organic acid mixture for 10 or more times and determining the amounts of metals removed by each extraction. The synthetic exudate formulated for the extractions was designed to mimic conditions of pH and organic acid composition and concentration in the rhizosphere. The data were fitted to the model

$$M_q = a\,(1 - e^{-bq}) \qquad \text{[Eq. 8.4-2]}$$

where M_q represents the cumulative metals removed (mg kg $^{-1}$) from soil that has been extracted cumulatively with q volumes of synthetic exudate (mol of COO^- kg $^{-1}$). In Equation 8.4-2, parameter a represents the ultimate metal removal (mg kg $^{-1}$). Therefore C in Equation 8.1-2 may be estimated by the value a derived from Equation 8.4-2. The k may be evaluated by parameter b (in kg per mol of COO^-), as

$$k = bA \qquad \text{[Eq. 8.6-6]}$$

For corn, the A values followed the probabilistic distribution described in Figure 8-13. Long-term field data from biosolids land application sites are needed to determine the A values for other plant species.

CHAPTER 9.0

SUMMARY AND CONCLUSIONS

The phytoavailability of metals in biosolids-treated soils may be described by the following model:

$$M = C\,(1 - e^{-kt})$$

where M (mg kg^{-1}) is the cumulative metal removal from the biosolids-treated soils by growing and harvesting plants for t years, C is the total phytoavailable metal pool (mg kg^{-1}), and k is the metal release coefficient (yr^{-1}) of the soils. If the C and k of a soil are defined, the phytoavailability of the metals in the biosolids receiving soils may be evaluated.

Organic acids present in the rhizosphere of growing plants are widely recognized to be responsible for dissolving the solid phase metals in the soil and making them available for plant absorption. Studies were initiated to investigate the concentration and composition of organic acids in the biosolids-treated plant growing media and the role of rhizosphere organic acids on the solubilization and phytoavailability of biosolids-borne metals. Based on the outcomes of the investigations, synthetic organic acid mixtures were formulated to assess the dissolution of metals from various biosolids sources in the biosolids-treated soils. Finally, soil and plant tissue samples from two long term biosolids land application sites were obtained to validate the metal phytoavailability model for the biosolids-treated soils with field data.

The findings are summarized as follows:

Literature Review. The nature of root exudates and their role in plant nutrition and soil chemistry has been studied for over two decades. Much of the root exudate literature consists of results from growing plant seedlings for up to 14 days under aseptic conditions. Studies have demonstrated that the primary component of root exudates are low molecular weight organic acids which might enhance the availability of sparingly soluble soil Fe, P, and Zn to growing plants. Microbial metabolism may play an important role in modifying the compositions of organic substances found in the rhizosphere. The composition of the organic acids in the root exudates, under such conditions varied considerably between the plant species studied. Even for the same plant species, investigators reported the presence of different chemical components. The environmental factors that affect the quantity of excretion and chemical composition of root exudates required further investigation.

Plant Growth. The biomass harvests of plants that were cultured with the biosolids-treated media were equal or better than those cultured with the standard plant nutrient solution media. The growth of plants was not adversely affected and plants exhibited no visual symptom of injuries even when the metal concentrations in plant tissue were significantly higher than normally observed. The increases in biomass production for plants that were associated with biosolids treatment were consistent over the entire experimental period (16 weeks) for all plant species tested (i.e., corn, wheat, Sudan grass, Swiss chard, canola, and chickpea), and for all biosolids sources (i.e., MWRDGC 1975, Milorganite, MWRDGC 1999, Los Angeles, and Riverside). The increased biomass yields for plants grown on biosolids-treated media over the standard nutrient solution was due to the presence of biosolids in the media that increased the intensity of availability of nutrients in the rhizosphere.

Root Exudates in Biosolids-treated Media. Organic acids were the primary substances found in root exudates that were recovered from all of the plant-growth media. Even the without-plant and the without-biosolids blanks produced organic acids. The biosolids treatment and the presence of plants enhanced the volume of organic acids produced. When plants were present in biosolids-treated media, the production of organic acids was further enhanced. Acetic, lactic, butyric, and oxalic acids were the primary components of organic acids found in the plant-growth media. Collectively, they accounted for 75-90% of the total organic acids recovered. Other organic acids detected were propionic, pyruvic, succinic, tartaric, maleic, and glutaric acids. The organic acid composition of the plant growing media did not change significantly with plant species, varieties of the same species, duration of plant growth, and sources of biosolids. The composition of recovered root exudates was independently confirmed by using the GC-MS and the NMR methods. The concentrations of organic acids in the plant growing media were estimated to be 0.001 to 0.1 M of COO^-, depending on the water content of the media. Based on the compositional analysis, a synthetic organic acid mixture—comparable to the composition of root exudates recovered from the rhizosphere of biosolids-treated media—was formulated to extract metals from biosolids-treated soils. The synthetic exudate contained 0.01 M COO^- , with a 13.5 mM $Ca(NO_3)_2$ background matrix and pH 4.8.

Metal Uptake by Plants. The concentrations of metals in plants grown on biosolids-treated media were significantly higher than those of the same plants grown on the standard plant nutrient media. The amounts of metal transferred from the biosolids-treated growing media to the plants varied with the metal elements. In descending order, the amounts transferred were: Cd > Ni = Zn > Cu = Pb > Cr. For corn grown on biosolids-treated media from seedlings to maturity, the amounts of metals absorbed by the plants were 14, 0.01, 1.6, 9.8, 1.4, and 8.6% of the biosolids-borne Cd, Cr, Cu, Ni, Pb, and Zn in the rhizosphere, respectively. For Cd, Cu, Ni, and Zn, the differences in metal uptake between plant species and varieties were relatively minor in comparison to the differences of metal uptake between plants grown on the standard plant nutrient solution- and the biosolids-treated media. The rate of metal uptake during the course of plant development may be calculated as the amounts of metal accumulated in plant tissue per unit biomass per unit growing period. The rate of Cd, and Zn uptake for corn shoots remained rather constant for the first 8-10 weeks, metabolically the most active period of growing season. Toward the end of the growing season, the uptake rates tended to rise as the biomass stopped accumulating while the uptake of water and metals continued. For Pb, the uptake rate by corn shoots grown on biosolids-treated media decreased with increasing length of the growing period. This result indicated that the solution concentration of Pb and therefore, its availability to the plant, is controlled by the rate of dissolution of biosolids-borne Pb.

Metal Solubility in Organic Acids. The solubility of metals in a biosolids-treated soil with 0.01 and 0.1 M synthetic organic acid mixtures was significantly higher than that at 0.001 M. The slight variations in the organic acid compositions due to biosolids treatment, length of growing periods, and plant species did not affect the solubility of the metals. The conclusion was reached by first using Student-Newman-Keuls test to compare amounts of metals extracted by standard media vs. biosolids-treated media. The differences were not significantly different at $p < 0.05$. The data then were pooled to compare the difference between plant growing periods using ANOVA. Again, the differences were not significant at $p < 0.05$. Results of this experiment indicated that the concentrations of synthetic organic acid mixtures significantly affected the amounts of metals extracted from the control as well as the biosolids-treated soils.

Metal Dissolution Kinetics. Based on the compositions of the organic acids in the recovered root exudates, synthetic root exudates were formulated to study the dissolution of metals by root exudates. Results of the experiments indicated that a rapid dissolution of metals occurred in the first 15 minutes of the equilibration. For Cd, Cu, Ni, and Zn, approximately 60–70% of the extractable metals were released in the first 15 minutes. For Cr and Pb, the initial releases were approximately 30% of the total. Following the initial release, a slow but steady release of metals occurred over the next 48 hours.

Field validation. Field validation was conducted using three biosolids application fields at the Fulton County Biosolids Land Reclamation Site of the Metropolitan Water Reclamation District of Greater Chicago and 16 experimental plots of long-term biosolids land application at the University of Minnesota.

The biosolids-treated soils from the testing plots were sequentially extracted with the synthetic organic acids for 10 consecutive times. The data showed that the cumulative metal released by the organic acids in the synthetic root exudates with respect to the volume of organic acids used followed the first order reaction model that: $M_q = a\,(1 - e^{-bq})$, where M_q is the cumulative metal removal from biosolids-treated soils (mg kg^{-1}) by q volume of organic acids (mole COO$^-$ kg^{-1}), a is the total synthetic root exudates extractable metal (mg kg^{-1}), and b is the synthetic root exudates metal release rate constant. The parameter a in this equation corresponds to the total phytoavailable metal pool (C) in the metal phytoavailability model.

The field sampling derived plant uptake data along with the estimates to the total phytoavailability metal pool of the soils then were used to fit the metal phytoavailability model to obtain the field-based metal plant uptake constant (k). Let A equal the effective organic acids production factor of the biosolids-treated soils. The k of a biosolids-treated soil may be mathematically related to the synthetic root exudates metal release constant of the corresponding soils as $k = b/A$, in which A followed a normal distribution with a mean of 3.8 and standard deviation of 1.2.

REFERENCES

Ae, N., J. Arihara, K. Okada, T. Yoshihara, and C. Johansen. 1990a. Phosphorus uptake by pigeon pea and its role in cropping systems of the Indian subcontinent. Science Washington 248:477-480.

Ae, N., J. Arihara, K. Okada, T. Yoshihara, and C. Johansen. 1990b. Uptake mechanism of iron-associated phosphorus in pigeonpea growing on Indian Alfisol and its significance to phosphorus availability in cropping systems. Pp. 164-169. Transactions 14th International Congress Soil Science, Kyoto, Japan. August 1990.

Ae, N., J. Arihara, K. Okada, T. Yoshihara, T. Otani, C. Johansen, P.J. Randall, E. Delhaize, R.A. Richards, and R. Munns. 1993. The role of piscidic acid secreted by pigeonpea roots grown in an Alfisol with low-P fertility. Genetic aspects of plant mineral nutrition: The Fourth International Symposium on Genetic Aspects of Plant Mineral Nutrition, 1991. Canberra, Australia. 279-288.

Aggarwal, K.K., J. Rajiv, and C.R. Babu. 1999. A rock-iron-solubilizing compound from root exudates of Tephrosia purpurea. J. Chem. Ecol. 25:2327-2336.

Ahumada, I., J. Mendoza, and L. Ascar. 1999. Determination of organic acids and phosphate in soil aqueous extracts by capillary zone electrophoresis. Commun. Soil Sci. Plant Anal. 30(1&2):213-220.

Ahumada, I.J. Mendoza, P. Escudero, K. Mossert, and L. Ascar. 2001. Determination of organic acids of low molecular weight and phosphate in soil by capillary electrophoresis. J. AOAC Int. 84:1057-1064.

Akers, H.A. 1983. Multiple hydroxamic acid microbial iron chelators (siderophores) in soils. Soil Sci. 135:156-159.

Ali, I.A., U. Kafkafi, I. Yamaguchi, Y. Sugimoto, and S. Inanaga. 2000. Growth, transpiration, root-borne cytokinins and gibberellins, and nutrient compositional changes in sesame exposed to low root-zone temperature under different ratios of nitrate: ammonium supply. J. Plant Nutr. 23:123-140.

Anderson, T.A., E.A. Guthrie, and B.T. Walton. 1993. Bioremediation in the rhizosphere. Environmental Science and Technology. 27:2630-2636.

Awad, F. and V. Romheld. 2000. Mobilization of heavy metals from a contaminated calcareous soil by plant borne and synthetic chelators and its uptake by wheat plants. J. Plant Nutr. 23:1847-1855.

Awad, F., V. Romheld, and H. Marschner. 1988. Mobilization of ferric iron from a calcareous soil by plant-borne chelators (phytosiderophores). J. Plant Nutr. 11:701-713.

Awad, F., V. Romheld, and H. Marschner. 1994. Effect of root exudates on mobilization in the rhizosphere and uptake of iron by wheat plants. Plant and Soil 165:213-218.

Azaizeh, H.A., H. Marschner, V. Romheld, and L. Wittenmayer. 1995. Effects of a vesicular-arbuscular mycorrhizal fungus and other soil microorganisms on growth, mineral nutrient acquisition and root exudation of soil-grown maize plants. Mycorrhiza 5:321-327.

Balke, N. 1985. Effects of allelochemicals on mineral uptake and associated physiological processes. The chemistry of allelopathy. ACS Symp. Ser. 268:161-178.

Barber, D.A. and K.B. Gunn. 1974. The effect of mechanical forces on the exudation of organic substances by the roots of cereal plants grown under sterile conditions. New Phytol. 73:39-45.

Barber, D.A. and J.K. Martin. 1976. The release of organic substances by cereal roots into soil. New Phytol. 76:69-80.

Barbour, W.M., D.R. Hattermann, and G. Stacey. 1991. Chemotaxis of Bradyrhizobium japonicum to soybean exudates. Appl. Environ. Microbiol. 57:2635-2639.

Bassam, B.J., M.A. Djordjevic, J.W. Redmond, M. Batley, and B.G. Rolfe. 1988. Identification of a nodD-dependent locus in the Rhizobium strain NGR234 activated by phenolic factors secreted by soybeans and other legumes. Molecular Plant Microbe Interactions 1:161-168.

Bastian, R.K. 1997. Biosolids management in the United States. Water Environment & Technology 9(5):45-50.

Bell, P.F., R.L. Chaney, and J.S. Angle. 1988. Staining localization of ferric reduction on roots. J. Plant Nutrition 11:1237-1252.

Blake, J.D., M.L. Clarke, and G.N. Richards. 1987. Determination of organic acids in sugar cane process juice by high performance liquid chromatography: improved resolution using dual Aminex HPX-87H cation-exchange columns equilibrated to different temperatures. J. Chromatography 398:265-277.

Bolan, N.S., M. J. Hedley, and R.E. White. 1989. Nitrogen fertilizer use: Fixation and soil acidification. p. 88-103. In R.E. White and L.D. Currie (ed.) Nitrogen in New Zealand agriculture and horticulture. Occasional Report No. 3. Fertilizer and Lime Research Centre, Massey University, Palmerston North, New Zealand.

Bolan, N.S., R. Naidu, S. Mahimairaja, and S. Baskaran. 1994. Influence of low-molecular-weight organic acids on the solubilization of phosphates. Biology and Fertility of Soils 18:311-319.

Booker, F.L., U. Blum, and E.L. Fiscus. 1992. Short-term effects of ferulic acid on ion uptake and water relations in cucumber seedlings. Journal of Experimental Botany 43:649-655.

Bozhkov, A.I., N.G. Menzyanova, and V.P. Leontovich. 1996. Lipid composition and antibacterial activity of root exudates secreted by wheat seedlings. Russian Journal of Plant Physiology 43:795-799.

Burken, J.G. and J.L. Schnoor. 1996. Phytoremediation: plant uptake of atrazine and role of root exudates. Journal of Environmental Engineering 122:958-963.

Cakmak, I. and H. Marschner. 1988a. Zinc-dependent changes in ESR signals, NADPH oxidase and plasma membrane permeability in cotton roots. Physiologia Plantarum 73:182-186.

Cakmak, I. and H. Marschner. 1988b. Increase in membrane permeability and exudation in roots of zinc deficient plants. Journal of Plant Physiology. 132:356-361.

California Association of Sanitation Agencies. 1999. Biosolids Management Practice Survey in the State of California, 1999 Update. Report prepared by Orange County Sanitation District, P.O. Box 8127, Fountain Valley, CA 92728. 21pp.

Cambier, P. and G. Sposito. 1991. Interactions of citric acid and synthetic hydroxy-aluminum montmorillonite. Clays Clay Miner. 39:158-166.

Candalaria, L.M. and A.C. Chang. 1997. Cadmium activities, solution speciation and solid phase distribution in cadmium nitrate and sewage sludge treated soil systems. Soil Sci. 162:722-732.

Chaboud, A. 1983. Isolation, purification and chemical composition of maize root cap slime. Plant and Soil 73:395-402.

Chaboud, A. and M. Rougier. 1991. Effect of root density in incubation medium on root exudate composition of axenic maize seedlings. Journal of Plant Physiology 137:602-606.

Chairidchai, P., and G.S.P. Ritchie. 1992. The effect of pH on zinc adsorption by a lateritic soil in the presence of citrate and oxalate. J. Soil Sci. 43:723-728.

Chairidchai, P. and G.S.P. Ritchie. 1993. Zinc adsorption by sterilized and non-sterilized soil in the presence of citrate and catechol. Communications in Soil Science and Plant Analysis 24:261-275.

Chaney, R.L. 1984. Diagnostic practices to identify iron deficiency in higher plants. Journal of Plant Nutrition 7:47-67.

Chaney, R.L. 1990. Twenty years of land application research. Biocycle 31:54-59.

Chaney, R.L. and J.A. Ryan. 1993. Toxic metals and toxic organic pollutants in MSW-composts: Research results on phyto-availability, bioavailability, fate, etc. p. 451-506. In H. A. Hoitink and H. M. Keener (ed.) Science and Engineering of Composting: Design, Environmental, Microbiological and Utilization Aspects. Renaissance Publications, Worthington, OH.

Chang, A.C., Hae-nam Hyun, and A.L. Page. 1997. Cadmium uptake for Swiss chard grown on composted sludge treated field plots: Plateau or time bomb? J. Environ. Qual. 26:11-19.

Chang, K.J. and J.K.M. Roberts 1991. Cytoplasmic malate levels in maize root-tips during K^+ ion uptake determined by 13C-NMR spectroscopy. Biochim. Biophys. Acta 1092:29-34.

Chhonkar, P.K. and J.C. Tarafdar. 1981. Characteristics and location of phosphatases in soil-plant system. Journal of the Indian Society of Soil Science 29:215-219.

Cieslinski, G., K.C.J. Van Rees, A.M. Szmigielska, and P.M. Huang. 1994. Plant growing conditions for collection of low molecular weight organic acids from root exudates. p. 264. In 1994 Agronomy Abstracts. ASA, Madison, WI.

Corey, R.B., P.A. Helmke, D.R. Keeney, G.C. Gerloff, R.T. Checkai. 1987a. EPA-600-2-87-002, U.S. EPA, Water Engineering Research laboratory, Cincinnati, OH. 322pp.

Corey, R.B., L.D. King, C. Lue-Hing, D.S. Fanning, J.J. Street, and J.M. Walker. 1987. Effects of Sludge Properties on Accumulation of Trace Elements by Crops. Pp. 25-52. In: Land Application of Sludge (Eds. A. L. Page, T. J. Logan, and J. A. Ryan). Lewis Publishers, Chelsea, MI. 168pp.

Curl, E.A. and B. Truelove. 1986a. Chapter 2, The structure and function of roots. In The rhizosphere. Springer-Verlag, Berlin, Germany. 9-54.

Curl, E.A. and B. Truelove. 1986b. Chapter 3, Root Exudates. In, The rhizosphere. Springer-Verlag, Berlin, Germany. 55-92.

Czarnes, S., P.D. Hallett, A.G. Bengough, and I.M. Young. 2000. Root- and microbial-derived mucilage affect soil structure and water transport. Eur. J. Soil Sci. 51:435-443.

Dakora, F.D., C.M. Joseph, and D.A. Phillips. 1993. Common bean root exudates contain elevated levels of daidzein and coumestrol in response to Rhizobium inoculation. Molecular Plant Microbe Interactions. 6:665-668.

D'Arcy-Lameta, A. 1986. Study of soybean and lentil root exudates. II. Identification of some polyphenolic compounds, relation with plantlet physiology. Plant and Soil 92:113-123.

Darrah, P.R. 1993. The rhizosphere and plant nutrition: A quantitative approach. Plant Soil 155/156: 1-20.

Delhaize, E., P.R. Ryan, and P.J. Randall. 1993. Aluminum tolerance in wheat (*Triticum aestivum* L.). II. Aluminum-stimulated excretion of malic acid from root apices. Plant Physiology 103:695-702.

Dinkelaker, B., V. Romheld, and H. Marschner. 1989. Citric acid excretion and precipitation of calcium citrate in the rhizosphere of white lupine (*Lupinus albus* L.). Plant, Cell and Environment 12:285-292.

Dixon, R.K., H.E. Garrett, and G.S. Cox. 1989. Boron fertilization, vesicular-arbuscular mycorrhizal colonization and growth of *Citrus jambhiri* Lush. Journal of Plant Nutrition 12:687-700.

Dowdy, R.H., W.E. Larson, J.M. Titrud, and J.J. Latterell. 1978. Growth and metal uptake of snap beans grown on sewage sludge-amended soil: a four-year field study. J. Environ. Qual. 7:252-257.

Eaton, F.M. 1936. Automatically operated sand-culture equipment. J. Agri. Res. 53: 433-444.

Elliott, H.A., M.R. Liberati, and C.P. Huang. 1986. Competitive adsorption of heavy metals by soils. J. Environ. Qual. 15:214-219.

Engels, C. and H. Marschner. 1993. Influence of the form of nitrogen supply on root uptake and translocation of cations in the xylem exudate of maize (*Zea mays* L.). Journal of Experimental Botany 44:1695-1701.

Evans, R.L. and J.J. Jurinak. 1976. Kinetics of phosphate release from a desert soil. Soil Sci. 121:205-211.

Fan, T.W.M., A.N. Lane, J. Pedler, D. Crowley, and R.M. Higashi. 1997. Comprehensive analysis of organic ligands in whole root exudates using nuclear magnetic resonance and gas chromatography mass spectrometry. Anal. Biochem. 251:57-68.

Firmin, J.L., K.E. Wilson, L. Rossen, and A.W.B. Johnston. 1986. Flavonoid activation of nodulation genes in Rhizobium reversed by other compounds present in plants. Nature. 324:90-92.

Fisher, W.R., H. Flessa, and G. Schaller. 1989. pH values and redox potentials in microsites of the rhizosphere. Z. Pflanzenernaehr. Bodenkd. 152:191-195.

Fox, T.R., N.B. Comerford, and W.W. McFee. 1990. Phosphorus and aluminium release from a spodic horizon mediated by organic acids. Soil Science Society of America Journal 54:1763-1767.

Foy, C.D., E.H. Lee, C.A. Coradetti, and G.J. Taylor. 1990. Organic acids related to differential aluminium tolerance in wheat (*Triticum aestivum*) cultivars. In, Plant nutrition physiology and applications. Van Beusichem, M.L., Ed. Kluwer Academic Publishers. Boston. 381-389.

Friebe, A., M. Schulz, P. Kuck, and H. Schnabl. 1995. Phytotoxins from shoot extracts and root exudates of Agropyron repens seedlings. Phytochemistry 38:1157-1159.

Furrer, G. and W. Stumm. 1986. The coordination chemistry of weathering. I. Dissolution kinetics of δ-Al2O3 and BeO. Geochim. Cosmochim. Acta 50:1847-1860.

Gahoonia, T.S. 1993. Influence of root-induced pH on the solubility of soil aluminium in the rhizosphere. Plant Soil 149:289-291.

Gardner, W.K., D.G. Parbery, D.A. Barber, and L. Swinden. 1983a. The acquisition of phosphorus by *Lupinus albus* L. V. The diffusion of exudates away from roots: a computer simulation. Plant and Soil 72:13-29.

Gardner, W.K., D.A. Barber, and D.G. Parbery. 1983b. The acquisition of phosphorus by *Lupinus albus* L. III. The probable mechanism by which phosphorus movement in the soil/root interfaces is enhanced. Plant and Soil 70:107-124.

Gardner, W.K., D.A. Barber, and D.G. Parbery. 1983c. Non-infecting rhizosphere micro-organisms and the mineral nutrition of temperate cereals. J. Plant Nutr. 6:185-199.

Gerke, J. 1994. Kinetics of soil phosphate desorption as affected by citric acid. Z. Pflanzenernaehr. Bodenkd. 157:17-22.

Gerritse, G.E., W. van Driel, K.W. Smilde, and B. van Luit. 1983. Uptake of heavy metals by crops in relation to their concentration in the soil solution. Plant and Soil 75:393-405.

Godo, G.H. and H.M. Reisenauer. 1980. Plant effects on soil manganese availability. J. Soil Sci. Soc. Am. 44:993-995.

Gogala, N. 1989. Growth substances in root exudate of *Pinus sylvestris* - their influence on mycorrhizal fungi. Agriculture, Ecosystems and Environment 28:151-154.

Gollany, H.T. and T.E. Schumacher. 1993. Combined use of colorimetric and microelectrode methods for evaluating rhizosphere pH. Plant Soil 154:151-159.

Graham, J.H., R.T. Leonard, and J.A. Menge. 1981. Membrane-mediated decrease in root exudation responsible for phosphorus inhibition of vesicular-arbuscular mycorrhiza formation. Plant Physiol. 68:548-552.

Graham, J.H., R.T. Leonard, and J.A. Menge. 1982. Interaction of light intensity and soil temperature with phosphorus inhibition of vesicular-arbuscular mycorrhiza formation. New Phytol. 91:683-690.

Gransee, A., and L. Wittenmayer. 2000. Qualitative and quantitative analysis of water-soluble root exudates in relation to plant species and development. J. Plant Nutr. Soil Sci. 163:381-385.

Gregg, P.E.H., A.D. Mackay, L.D. Currie, and J.K. Syers. 1988. Application strategies for Sechura phosphate rock use on permanent pastures. Fert. Res. 17:219-243.

Groleau-Renaud, V., S. Plantureux, A. Tubeileh, and A. Guckert. 2000. Influence of microflora and composition of root bathing solution on root exudation of maize plants. J. Plant Nutr. 23:1283-1301.

Habib, L., C. Chenu, J.L. Morel, and A. Guckert. 1990. Adsorption of maize root mucilages on homoionic clays - consequences on the microstructure of the complexes. Comptes Rendus de L Academie des Sciences Serie II - Mecanique Physique Chimie Sciences de L'univers Sciences de La Terre 310:1541-1546.

Hale, M.G. and L.D. Moore. 1979. Factors affecting root exudation. II. 1970-1978. Advances in Agronomy 31:93-124.

Hale, M.G., C.L. Foy, and F.J. Shay. 1971. Factors affecting root exudation. Adv. Agron. 23:89-109.

Hale, M.G., L.D. Moore, and G.J. Griffin. 1978. Root exudates and exudation. p. 163-203. In: Y.R. Demmergues and S.V. Krupa (ed.) Interactions between nonpathogenic soil microorganisms and plants. Elsevier Science Publishers, New York.

Haller, T. and H. Stolp. 1985. Quantitative estimation of root exudation of maize plants. Plant and Soil 86:207-216.

Halm, B.J., J.W. Stewart, and R.L. Halstead. 1972. The phosphorus cycle in a native grassland ecosystem. In: Symposium on the Use of isotopes and radiation in soil plant relations including forestry. International Atomic Energy Agency, Vienna. 571-586.

Harper, J.R. and N.E. Balke. 1981. Characterization of the inhibition of K^+ potassium ion absorption in oat roots by salicylic acid. Plant Physiol. 68:1349-1353.

Harter, R.D. 1983. Effect of pH on adsorption of lead, copper, zinc, and nickel. Soil Sci. Soc. Am. J. 47:47-51.

Helal, H.M. and A. Dressler. 1989. Mobilization and turnover of soil phosphorus in the rhizosphere. Z. Pflanz. Bodenk. 152:175-180.

Helal, H.M. and D.R. Sauerbeck. 1984. Influence of plant roots on C and P metabolism in soil. Plant and Soil. 76:175-182.

Hether, N.H., R.A. Olsen, and L.L. Jackson. 1984. Chemical identification of iron reductants exuded by plant roots. Journal of Plant Nutrition. 7:667-676.

Hinsinger, P., and R.J. Gilkes. 1996. Mobilization of phosphate from phosphate rock and alumina-sorbed phosphate by the roots of ryegrass and clover as related to rhizosphere pH. Eur. J. Soil Sci. 47:533-544.

Hodge, A., S.J. Grayston, and B.G. Ord. 1996. A novel method for characterisation and quantification of plant root exudates. Plant and Soil 184:97-104.

Hoffland, E., G. Findenegg, and J. Nelemans. 1989. Solubilization of rock phosphate by rape. II. Local root exudation of organic acids as a response to P-starvation. Plant and Soil 113:161-165.

Horvath, B., E. Kondorosi, M. John, J. Schmidt, Torok, I, Z. Gyorgypal, Barabas, I, U. Wieneke, J. Schell, and A. Kondorosi. 1986. Organization, structure and symbiotic function of *Rhizobium meliloti* nodulation genes determining host specificity for alfalfa. Cell 46:335-343.

Hyun, Hae-nam, A.C. Chang, D.R. Parker, A.L. Page. 1998. Cadmium solubility and phytoavailability in sludge-treated soil: Effects of soil organic carbon. J. Environ. Qual. 27:329-334.

Iijima, M., B. Griffiths, and A.G. Bengough. 2000. Sloughing of cap cells and carbon exudation from maize seedling roots in compacted sand. New Phytol. 145:477-482.

Inskeep, W.P. and S.D. Comfort. 1986. Thermodynamic predictions for the effects of root exudates on metal speciation in the rhizosphere. Journal of Plant Nutrition 9:567-586.

Janssen, R.P.T., M.G.M. Bruggenwert, and W.H. Van Riemsdijk. 1997. Interaction between citrate and montmorillonite-Al hydroxide polymer system. Eur. J. Soil Sci. 48:463-472.

Jauregui, M.A. and H.M. Reisenauer. 1982. Dissolution of oxides of manganese and iron by root exudate components. Soil Sci. Soc. Am. J. 46:314-317.

Jenny, H. and R. Overstreet. 1938. Contact effects between plant roots and soil colloids. Proc. National Academy of Science 24:384-392.

Jing, J., and T.J. Logan. 1992. Effects of sewage sludge cadmium concentration on chemical extractability and plant uptake. J. Environ. Qual. 21:73-81.

Jones, D.L. and P.R. Darrah. 1993a. Influx and efflux of amino acids from Zea mays L. roots and their implications for N nutrition and the rhizosphere. Plant nutrition from genetic engineering to field practice: Proceedings of the Twelfth International Plant Nutrition Colloquium, September 1993, Perth, Western Australia. September. 91-94.

Jones, D.L. and P.R. Darrah. 1993b. Re-sorption of organic compounds by roots of Zea mays L. and its consequences in the rhizosphere. II. Experimental and model evidence for simultaneous exudation and re-sorption of soluble C compounds. Plant and Soil. 153:47-59.

Jones, D.L., and P.R. Darrah. 1994. Amino-acid influx at the soil-root interface of Zea mays L. and its implications in the rhizosphere. Plant Soil 163:1-12.

Jones, D.L., and P.R. Darrah. 1996. Re-sorption of organic-compounds by roots of Zea mays L. and its consequences in the rhizosphere. 3. Characteristics of sugar influx and efflux. Plant Soil 178:153-160.

Jones, D.L., and L.V. Kochian. 1996. Aluminum-organic acid interactions in acid soils. 1. Effect of root-derived organic-acids on the kinetics of Al dissolution. Plant Soil 182:221-228.

Jones, D.L., A.C. Edwards, K. Donachie, and P.R. Darrah. 1994. Role of proteinaceous amino acids released in root exudates in nutrient acquisition from the rhizosphere. Plant and Soil. 158:183-192.

Jones, D.L., A.M. Prabowo, and L.V. Kochian. 1996a. Kinetics of malate transport and decomposition in acid soils and isolated bacterial-populations – the effect of microorganisms on root exudation of malate under Al stress. Plant Soil 182:239-247.

Jones, D.L., P.R. Darrah, and L.V. Kochian. 1996b. Critical evaluation of organic acid mediated iron dissolution in the rhizosphere and its potential role in root iron uptake. Plant Soil 180:57-66.

Kabata-Pendias. 2001. Trace Elements in Soils and Plants, 3rd Edition. CRC Press, Boca Raton, Florida. 413pp.

Kato-Noguchi, H., J. Mizutani, and K. Hasegawa. 1994. Allelopathy of oats. II. Allelochemical effect of L-tryptophan and its concentration in oat root exudates. Journal of Chemical Ecology 20:315-319.

Knight, B., F.J. Zhao, S.P. McGrath, and Z.G. Shen. 1997. Zinc and cadmium uptake by the hyperaccumulator *Thlaspi caerulescens* in contaminated soils and its effects on the concentration and chemical speciation of metals in soil solution. Plant Soil 197:71-78.

Kobayashi, A., Kim-MyongJo, K. Kawazu, and M.J. Kim. 1996. Uptake and exudation of phenolic compounds by wheat and antimicrobial components of the root exudate. Zeitschrift fur Naturforschung. 51:527-533.

Kothari, S.K., H. Marschner, and V. Romheld. 1991. Effect of a vesicular-arbuscular mycorrhizal fungus and rhizosphere micro-organisms on manganese reduction in the rhizosphere and manganese concentrations in maize (*Zea mays* L.). New PhytologistKozdroj, J., and J.D. Van Elsas. 2000. Response of the bacterial community to root exudates in soil polluted with heavy metals assessed by molecular and cultural approaches. Soil Biol. Biochem. 32:1405-1417.

Kraffczyk, I., G. Trolldenier, and H. Beringer. 1984. Soluble root exudates of maize: influence of potassium supply and rhizosphere microorganisms. Soil Biology and Biochemistry 16:315-322.

Krishnamurti, G.S.R., G. Cieslinski, P.M. Huang, and K.C.J. Van Rees. 1997. Kinetics of cadmium release from soils as influenced by organic acids: Implication in cadmium availability. J. Environ. Qual. 26:271-277.

Krzyszowska, A.J., M.J. Blaylock, G.F. Vance, and M.B. David. 1996. Ion-chromatographic analysis of low molecular weight organic acids in spodosol forest floor solutions. Soil Sci. Soc. Am. J. 60:1565-1571.

Laheurte, F., C. Leyval, and J. Berthelin. 1990. Root exudates of maize, pine and beech seedlings influenced by mycorrhizal and bacterial inoculation. Symbiosis 9:111-116.

Li, M.G., T. Shinano, and T. Tadano. 1997. Distribution of exudates of lupin roots in the rhizosphere under phosphorus deficient conditions. Soil Sci. Plant Nutr. 43:237-245.

Lilieholm, B.C., L.M. Dudley, and J.J. Jurinak. 1992. Oxalate determination in soils using ion chromatography. Soil Science Society of America Journal 56:324-326.

Linehan, D.J., A.H. Sinclair, and M.C. Mitchell. 1989. Seasonal changes in Cu, Mn, Zn and Co concentrations in soil in the root-zone of barley (*Hordeum vulgare* L.). Journal of Soil Science 40:103-115.

Lipton, D.S., R.W. Blanchar, and D.G. Blevins. 1987. Citrate, malate, and succinate concentration in exudates from P- sufficient and P-stressed *Medicago sativa* L. seedlings. Plant Physiology 85:315-317.

Loeppert, R.H. 1985. Chapter 19. Chemistry of iron in calcareous systems. In: Iron in Soils and Clay Minerals. Stucki J.W., B.A. Goodman, and U. Schwertmann (Eds.). Reidel Publishing Company. Dordrecht, Netherlands. 689-713.

Loeppert, R.H. and C.T. Hallmark. 1985. Indigenous soil properties influencing the availability of iron in calcareous soil. Soil Science Society of America Journal 49:597-603.

Loeppert, R.H., S.C. Geiger, R.C. Hartwig, and D.R. Morris. 1988. A comparison of indigenous soil factors influencing the Fe-deficiency chlorosis of sorghum and soybean in calcareous soils. Journal of Plant Nutrition 11:1481-1492.

Logan, T.J. and R.L. Chaney. 1983. Utilization of municipal wastewater and sludge on land—metals. p. 235-295. In: A. L. Page et al. (ed.) Utilization of municipal wastewater and sludge on land. University of California, Riverside, CA.

Lynch, J.M. 1982. The rhizosphere zone of bacterial growth which was stimulated around the roots of legumes. Pp. 395-411. In: G. Richard (ed.), Experimental Microbial Ecology.

Lyu, S.W. and U. Blum. 1990. Effects of ferulic acid, an allelopathic compound, on net P, K, and water uptake by cucumber seedlings in a split-root system. Journal of Chemical Ecology 16:2429-2439.

Ma, J.F. and K. Nomoto. 1993. Inhibition of mugineic acid-ferric complex uptake in barley by copper, zinc and cobalt. Physiologia Plantarum 89:331-334.

Marschner, H. (ed.) 1995. Mineral nutrition of higher plants, 2nd ed. Academic Press. London, UK.

Marschner, H. 1998. Role of root growth, arbuscular mycorrhiza, and root exudates for the efficiency in nutrient acquisition. Field Crop Res. 56:203-207.

Marschner, H. and V. Romheld. 1983. In vivo measurement of root-induced pH changes at the soil-root interfaces: Effect of plant species and nitrogen source. Z. Pflanzenernaehr. Bodenkd. 111:241-251.

Marschner, H., V. Romheld, and H. Ossenberg-Neuhaus. 1982. Rapid method for measuring changes in pH and reducing processes along roots of intact plants. Z. Pflanzenernaehr. Bodenkd. 105:407-416.

Marschner, H., Romheld, V, and M. Kissel. 1986. Different strategies in higher plants in mobilization and uptake of iron. Journal of Plant Nutrition. 9:695-713.

Marschner, H., V. Romheld, and I. Cakmak. 1987. Root-induced changes of nutrient availability in the rhizosphere. J. Plant Nutr. 10:1175-1184.

Marschner, P., D.E. Crowley, and R.M. Higashi. 1997. Root exudation and physiological status of a root-colonizing fluorescent pseudomonad in mycorrhizal and non-mycorrhizal pepper (Capsicum annuum L.). Plant and Soil 189:11-20.

Martell, A.E., and R.M. Smith (eds.). 1976-1989. Critical stability constants, 6th ed. Plenum Press, New York.

Martin, J.K. 1977. Factors influencing the loss of organic carbon from wheat roots [Soils]. Soil Biol. Biochem. 9:1-7.

Masaoka, Y., M. Kojima, S. Sugihara, T. Yoshihara, M. Koshino, A. Ichihara, and N.J. Barrow. 1993. Dissolution of ferric phosphate by alfalfa (*Medicago sativa* L.) root exudates. Plant and Soil 155/156:75-78.

Mathews, A., R.M. Kosslak, C. Sengupta-Gopalan, E.R. Appelbaum, B.J. Carroll, and P.M. Gresshoff. 1989. Biological characterization of root exudates and extracts from nonnodulating and supernodulating soybean mutants. Molecular Plant-Microbe Interactions 2:283-290.

McFeeters, R.F. 1993. Single-injection HPLC analysis of acids, sugars, and alcohols in cucumber fermentations. J. Agr. Food Chem. 41:1439-1443.

McLaughlin, M.J. and A.M. Alston. 1986. Measurement of phosphorus in the soil microbial biomass: a modified procedure for field soils. Soil Biology and Biochemistry 18:437-443.

Meharg, A.A. and K. Killham. 1990. The effect of soil pH on rhizosphere carbon flow of *Lolium perenne*. Plant and Soil 123:1-7.

Meharg, A.A. and K. Killham. 1991. A novel method of quantifying root exudation in the presence of soil microflora. Plant and Soil 133:111-116.

Meharg, A.A. and K. Killham. 1995. Loss of exudates from the roots of perennial ryegrass inoculated with a range of micro-organisms. Plant and Soil 170:345-349.

Mench, M. and E. Martin. 1991. Mobilization of cadmium and other metals from two soils by root exudates of *Zea mays* L., *Nicotiana tabacum* L. and *Nicotiana rustica* L. Plant and Soil 132:187-196.

Mench, M., J.L. Morel, A. Guckert, and B. Guillet. 1988a. Metal binding with root exudates of low molecular weight. Journal of Soil Science 39:521-527.

Mench, M., J.L. Morel, and A. Guckert. 1988b. Influence of metal [Cd (II), Cu (II), Pb (II), Zn (II)] ions on soluble root exudates of maize (*Zea mays* L.). Agronomie 8:237-241.

Merckx, R., J.H. Ginkel, J. Sinnaeve, and A. Cremers. 1986. Plant-induced changes in the rhizosphere of maize and wheat. II. Complexation of cobalt, zinc and manganese in the rhizosphere of maize and wheat. Plant and Soil 96:95-107.

Mergel, A.A., A.V. Timchenko, V.A. Leoshko, and V.N. Kudeyarov. 1992. Participation of nitrogen released from roots in nitrogen transformation in soil and in the process of formation of soil extra-nitrogen. Agrokhimiya 3-13.

Mergel, A.A., A.V. Timchenko, and V.N. Kudeyarov. 1996. The role of root exudates in transformation of nitrogen- and carbon-bound compounds in soils. Eurasian Soil Science 29:1151-1155.

Milward, C.G. and P.D. Klukner. 1989. Microwave digestion technique for the extraction of minerals from environmental marine samples for analysis by inductively coupled plasma emission spectrometry and atomic absorption spectrometry. J. Anal. Atomic Spectrosc. 4:709-713.

Misra, A. and S. Ramani. 1992. Identification of caffeic acid in the root exudate of Fe-efficient Japanese mint subjected to Fe stress: its action on Fe-chlorosis recovery. Agrochimica 36:93-100.

Miwa, H. 1985. Reversed-phase ion-pair chromatography of some biologically important carboxylic acids as 2-nitrophenylhydrazides. Journal of Chromatography 333:215-219.

Morel, J.L., M. Mench, and A. Guckert. 1986. Measurement of Pb2+, Cu2+, Cd2+ binding with mucilage exudates from maize (Zea mays L.) roots. Biol. Fertil. Soils. 2:29-34.

Morel, J.L., M. Mench, and A. Guckert. 1987. Heavy metal dynamics in the rhizosphere: role of root exudates. Revue d'Ecologie et de Biologie du Sol 24:485-492.

Mori, S., N. Nishizawa, S. Kawai, Y. Sato, and S. Takagi. 1987. Dynamic state of mugineic acid and analogous phytosiderophores in Fe-deficient barley. Journal of Plant Nutrition 10:1003-1011.

Morse, C.C., I.V. Yevdokimov, and T.H. DeLuca. 2000. In situ extraction of rhizosphere organic compounds from contrasting plant communities. Commun. Soil Sci. Plant Anal. 31:725-742.

Mozarfar, A. 1989. Iron-reducing root exudates: a modified microanalytical method. Journal of Plant Nutrition 12:1215-1229.

Muller, S., A. Merwe, van-der, H. Schildknecht, J.H. Visser, and A. Van-der-Merwe. 1993. An automated system for large-scale recovery of germination stimulants and other root exudates. Weed Science. 41:138-143.

Mullette, K.J., N.J. Hannon, and A.G.L. Elliott. 1974. Insoluble phosphorus usage by eucalyptus. Plant and Soil 41:199-205.

Murakami, T., K. Ise, M. Hayakawa, S. Kamei, and S.I. Takagi. 1989. Stabilities of metal complexes of mugineic acids and their specific affinities for iron(III). Chem. Lett. 2137-2140.

Muranyi, A., B. Seeling, E. Ladewig, and A. Jungk. 1994. Acidification of the rhizosphere of rape seedlings by nitrification and ammonium uptake. Z. Pflanzenernaehr. Bodenkd. 157:61-65.

Nardi, S., G. Dell'Agnola, and A. Albuzio. 1986. Changes in the molecular complexity of humus caused by organic acids. Agrochimica 30:148-159.

Nardi, S., F. Reniero, and G. Concheri. 1997. Soil organic matter mobilization by root exudates of three maize hybrids. Chemosphere. 35:2237-2244.

Nardi, S., G. Concheri, D. Pizzeghello, A. Sturaro, R. Rella, and G. Parvoli. 2000. Soil organic matter mobilization by root exudates. Chemosphere 41:653-658.

Nielsen, N.E. 1976. The effect of plants on the copper concentration in the soil solution. Plant Soil 45:679-687.

Nishizawa, N. and S. Mori. 1987. The particular vesicle appearing in barley root cells and its relation to mugineic acid secretion. Journal of Plant Nutrition 10:1013-1020.

Nyatsanga, T., and W.H. Pierre. 1973. Effect of nitrogen fixation by legumes on soil acidity. Agron. J. 65:936-940.

Nye, P.H. 1981. Changes of pH across the rhizosphere induced by roots. Plant Soil 61: 7-26.

Nye, P.H., D. Vaughan, and A.C. Edwards. 1977. Movement and uptake of solutes in field conditions. In: Solute Movement on the Soil Root System. Studies in Ecology. Blackwell, Oxford. 243-323.

Nye, P.H., W. Smith, H., R. Adams, E., J. Allison, R., J. Irvine, E.N. Knoepfler. 1984. On estimating the uptake of nutrients solubilized near roots or other surfaces. Journal of Soil Science 35:439-446.

Ochs, M., I. Brunner, W. Stumm, and B. Cosovic. 1993. Effects of root exudates and humic substances on weathering kinetics. Water, Air, and Soil Pollution 68:213-229.

Olsen, R.A. and J.C. Brown. 1980. Factors related to iron uptake by dicotyledonous and monocotyledonous plants. II. The reduction of Fe3+ iron ion as influenced by roots and inhibitors. J. Plant Nutr. 2:647-660.

Olsen, R.A., J.H. Bennett, D. Blume, and J.C. Brown. 1981. Chemical aspects of iron stress response mechanism in tomatoes. J. Plant Nutr. 3:905-921.

Osmond, C.B. 1976. Ion absorption and carbon metabolism in cells of higher plants. p. 347-372. In: U. Liittge and M.G. Pitman (ed.), Encyclopedia of Plant Physiology. New Series, Vol. 2, Part A. Springer-Verlag, Berlin.

Otani, T., N. Ae, and H. Tanaka. 1996. Phosphorus (P) uptake mechanisms of crops grown in soils with low P status. Soil Science and Plant Nutrition 42:553-560.

Pant, H.K., D. Vaughan, and A.C. Edwards. 1994. Molecular size distribution and enzymatic degradation of organic phosphorus in root exudates of spring barley. Biology and Fertility of Soils 18:285-290.

Parfitt, R.L. 1978. Anion adsorption by soils and soil minerals. Adv. Agron. 30:1-50.

Parfitt, R.L. 1979. The availability of P from phosphate-goethite bridging complexes: Description and uptake by ryegrass. Plant Soil 53:55-65.

Parker, D.R., R.L. Chaney, and W.A. Norvell. 1995. Chemical equilibrium models: applications to plant nutrition research. pp 163-200 In: R. H. Lippert (ed.), Soil Chemical Equilibria and Reaction Models. Soil Soc. Soc. Amer. Special Publication No. 42. Madison, WI.

Perez-Alfocea, F., M.E. Balibrea, J.J. Alarcon, and M.C. Bolarin. 2000. Composition of xylem and phloem exudates in relation to the salt-tolerance of domestic and wild tomato species. J. Plant Physiol. 156:367-374.

Petersen, W. and M. Bottger. 1991. Contribution of organic acids to the acidification of the rhizosphere of maize seedlings. Plant and Soil 132:159-163.

Piccolo, A., S. Nardi, and G. Concheri. 1992. Structural characteristics of humic substances as related to nitrate uptake and growth regulation in plant systems. Soil Biology and Biochemistry 24:373-380.

Pohlman, A.A., and J.G. McColl. 1986. Kinetics of metal dissolution from forest soils by soluble organic acids. J. Environ. Qual. 15:86-92.

Pojasok, T. and B.D. Kay. 1990. Effect of root exudates from corn and bromegrass on soil structural stability. Canadian Journal of Soil Science 70:351-362.

Pomilio, A.B., S.R. Leicach, M.Y. Grass, C.M. Ghersa, M. Santoro, and A.A. Vitale. 2000. Constituents of the root exudates of *Avena fatua* grown under far-Infrared-enriched light. Phytochem. Anal. 11:304-308.

Porter, J.R. 1986. Production of a putative phytosiderophore by soybeans in response to iron deficiency stress. Journal of Plant Nutrition 9:1113-1121.

Powell, P.E., P.J. Szaniszlo, G.R. Cline, and C.P.P. Reid. 1982. Hydroxamate siderophores in the iron nutrition of plants. Journal of Plant Nutrition 5:653-673.

Pramanik, M.H.R., M. Nagai, T. Asao, and Y. Matsui. 2000. Effects of temperature and photoperiod on phytotoxic root exudates of cucumber (*Cucumis sativus*) in hydroponic culture. J. Chem. Ecol. 26:1953-1967.

Prikryl, Z. and V. Vancura. 1980. TI: Root exudates of plants. VI. Wheat root exudation as dependent on growth, concentration gradient of exudates and the presence of bacteria *Pseudomonas putida*. Plant and Soil 57:69-83.

Rengel, Z., and V. Romheld. 2000. Root exudation and Fe uptake and transport in wheat genotypes differing in tolerance to Zn deficiency. Plant Soil 222:25-34.

Riley, D., and S.A. Barber. 1971. Effect of ammonium and nitrate fertilization on phosphorus uptake as related to root-induced pH changes at the root-soil interface. Soil Sci. Soc. Am. Proc. 35:301-306.

Romheld, V., and F. Awad. 2000. Significance of root exudates in acquisition of heavy metals from a contaminated calcareous soil by graminaceous species. J. Plant Nutr. 23:1857-1866.

Romheld, V. and H. Marschner. 1983. Mechanism of iron uptake by peanut plants. I. Fe(III) reduction, chelate splitting, and release of phenolics. Plant Physiology 71:949-954.

Romheld, V., and H. Marschner. 1984. Plant-induced pH changes in the rhizosphere of 'Fe-efficient' and 'Fe-inefficient' soybean and corn cultivars. J. Plant Nutr. 7:623-630.

Romheld, V. and H. Marschner. 1986. Evidence for a specific uptake system for iron phytosiderophores in roots of grasses. Plant Physiology 80:175-180.

Romheld, V. and H. Marschner. 1990. Genotypical differences among graminaceous species in release of phytosiderophores and uptake of iron phytosiderophores. Plant and Soil 123:147-153.

Rovira, A.D. 1969. Plant root exudates. Bot. Rev. 35:35-57.

Rovira, A.D. 1979. Biology of the soil-root interface. In: The Soil-Root Interface. Harley J.L. and R.S. Russell (eds.). Academic Press. London, UK. 145-160.

Rovira, A.D. and C.B. Davey. 1974. Biology of the rhizosphere. In: The plant root and its environment. Carson, E.W. (ed.), University of Virginia Press, Charlottesville. 153-204.

Rovira, A.D., G.D. Bowen, and R.C. Foster. 1983. The significance of rhizosphere microflora and mycorrhizas in plant nutrition. In: Inorganic Plant Nutrition. Lauchli A. and R.L. Bieleski (eds.). Spring-Verlag, Berlin. 61-93.

Rudolfs, W. and H.W. Gehm. 1942. Chemical Composition of Sewage Sludges, with Particular Reference to Their Phosphoric Acid Contents. Bulletin 699, New Jersey Agricultural Experiment Station, Rutgers University, New Brunswick, NJ.

Ruiz, L. and J.C. Arvieu. 1990. Measurement of pH gradients in the rhizosphere. Symbiosis 9:71-75.

Saeki, Y., T. Yamakawa, M. Ikeda, and J. Ishizuka. 1996. Effects of root exudates of Rj2, Rj3- and Rj4-genotype soybeans on growth and chemotaxis of *Bradyrhizobium japonicum*. Soil Science and Plant Nutrition 42:413-417.

Sakurai, K., and P.M. Huang. 1995. Cd adsorption on the hydroxyaluminum-montmorillonite complex as influenced by oxalate. p. 39-46. In: P.M. Huang et al. (ed.), Environmental Impact of Soil Component Interactions. Lewis Publisher, FL.

Schilling, G., A. Gransee, A. Deubel, G. Lezovic, and S. Ruppel. 1998. Phosphorus availability, root exudates, and microbial activity in the rhizosphere. Z. Pflanz. Bodenk. 161:465-478.

Schonwitz, R. and H. Ziegler. 1982. Exudation of water-soluble vitamins and of some carbohydrates by intact roots of maize seedlings (*Zea mays* L.) into a mineral nutrient solution. Z. Pflanzenphysiol. Int. J. Plant Physiol. 107:7-14.

Schurr, U. and E.D. Schulze. 1995. The concentration of xylem sap constituents in root exudate, and in sap from intact, transpiring castor bean plants (*Ricinus communis* L.). Plant, Cell and Environment 18:409-420.

Shen, Y., L. Strom, J.A. Jonsson, and G. Tyler. 1996. Low-molecular organic-acids in the rhizosphere soil solution of beech forest (*Fagus sylvatica* L.) cambisols determined by ion chromatography using supported liquid membrane enrichment technique. Soil Biol. Biochem. 28:1163-1169.

Siciliano, S.D., H. Goldie, and J.J. Germida. 1998. Enzymatic activity in root exudates of Dahurian wild rye (*Elymus dauricus*) that degrades 2-chlorobenzoic acid. J. Agr. Food Chem. 46:5-7.

Smith, W.H. 1970. Root exudates of seedling and mature sugar maple. Phytopathology 60:701-703.

Soejima, H., T. Sugiyama, and K. Ishihara. 1992. Changes in cytokinin activities and mass spectrometric analysis of cytokinins in root exudates of rice plants (*Oryza sativa* L.). Comparison between cultivars Nipponbare and Akenohoshi. Plant Physiology 100:1724-1729.

Sommers, L., V. Van Volk, P.M. Giordano, W.E. Sopper, and R. Bastian. 1987. Effects of Soil Properties on Accumulation of Trace Elements by Crops. Pp. 5-24. In: Land Application of Sludge (Eds. A.L. Page, T.J. Logan, and J.A. Ryan). Lewis Publishers, Chelsea, MI. 168pp.

Sommers, L.E., A.L. Page, T.J. Logan, and J.A. Ryan. 1991. Impact of Sewage Sludge on Soils and Barley: A Regional Study. USDA-CSRS Regional Research Publication W-124: Optimum Use of Sewage Sludge on Agricultural Land. Agricultural Experiment Station, Colorado State University.

Sparks, D.L. 1984. Ion activities: An historical and theoretical overview. Soil Sci. Soc. Amer. J. 48:415-418.

Sposito, G. 1981. Trace metals in contaminated waters. Environ. Sci. Tech. 15:396-403.

Steele, H.L., D. Werner, and J.E. Cooper. 1999. Flavonoids in seed and root exudates of *Lotus pedunculatus* and their biotransformation by *Mesorhizobium loti*. Physiol. Plant. 107:251-258.

Stevenson, F.J. (ed.) 1986. Cycles of Soil: Carbon, nitrogen, phosphorus, sulfur, and micronutrients. John Wiley & Sons, Inc., New York.

Stumm, W., and J.J. Morgan. (ed.) 1996. Aquatic chemistry: Chemical equilibria and rates in natural waters. 3rd ed. John Wiley & Sons, Inc., New York.

Subbarao, G.V., N. Ae, and T. Otani. 1997. Genotypic variation in iron-, and aluminum-phosphate solubilizing activity of pigeonpea root exudates under P deficient conditions. Soil Sci. Plant Nutr. 43:295-305.

Swinnen, J., J.A. Veen, van, R. Merckx, and J.A. Van-Veen. 1994. Rhizosphere carbon fluxes in field-grown spring wheat: model calculations based on ^{14}C partitioning after pulse-labeling. Soil Biology and Biochemistry 26:171-182.

Szmigielska, A.M., K.C.J. Vanrees, G. Cieslinski, P.M. Huang, and D.R. Knott. 1995. Determination of low molecular weight dicarboxylic acids in root exudates by gas chromatography. J. Agr. Food Chem. 43:956-959.

Tadano, T. and H. Sakai. 1991. Secretion of acid phosphatase by the roots of several crop species under phosphorus-deficient conditions. Soil Sci. Plant Nutr. 37:129-140.

Tagliavini, M., A. Masia, and M. Quartieri. 1995. Bulk soil pH and rhizosphere pH of peach trees in calcareous and alkaline soils as affected by the form of nitrogen fertilizers. Plant Soil 176:263-271.

Takagi, S., S. Kamei, and M.H. Yu. 1988. Efficiency of iron extraction from soil by mugineic acid family phytosiderophores. Journal of Plant Nutrition 11:643-651.

Tang, C.S. and C.C. Young. 1982. Collection and identification of allelopathic compounds from the undisturbed root system of bigalta limpograss (*Hemarthria altissima*). Plant Physiol. 69:155-160.

Taniguchi, S., N. Yamagata, and K. Sakurai. 2000. Cadmium adsorption on hydroxyaluminosilicate-montmorillonite complex as influenced by oxalate and citrate. Soil Sci. Plant Nutr. 46:315-324.

Tate, K.R. 1985. Soil phosphorus. In: Soil organic matter and biological activity. Vaughan, D. Ed. Boston, Martinus Nijhoff, Dordrecht. 329-377.

Traina, S.J., G. Sposito, D. Hesterberg, and U. Kafkafi. 1986. Effects of pH and organic acids on orthophosphate solubility in an acid, montmorillonitic soil. Soil Science Society of America Journal 50:45-52.

Traore, O., V. Groleau-Renaud, S. Plantureux, A. Tubeileh, V. Boeuf-Tremblay. 2000. Effect of root mucilage and modeled root exudates on soil structure. Eur. J. Soil Sci. 51:575-581.

Treeby, M., H. Marschner, and V. Romheld. 1989. Mobilization of iron and other micronutrient cations from a calcareous soil by plant-borne, microbial, and synthetic metal chelators. Plant and Soil 114:217-226.

Trofymow, J., A.D. Coleman, and C. Cambardella. 1987. Rates of rhizodeposition and ammonium depletion in the rhizosphere of axenic oats roots. Plant and Soil 97:333-344.

Unger, M.T. and H.E. Allen. 1988. In: Heavy metals in the hydrological cycle. M. Astruc and J.N. Lester (ed.). Selper, Londen, pp. 481-488.

Uren, N.C. and H.M. Reisenauer. 1988. The role of root exudates in nutrient acquisition. Advances in Plant Nutrition [edited by Tinker, B.]. 3:79-114.

Uselman, S.M., R.G. Qualls, and R.B. Thomas. 2000. Effects of increase atmospheric CO2, temperature, and soil N availability on root exudation of dissolved organic carbon by a N-fixing tree (*Robinia pseudoacacia* L.). Plant Soil 222:191-202.

U.S. EPA. 1993. Standards for the Use or Disposal of Sewage Sludge; Final Rules. Fed. Reg. 58:9248-9404.

Vancura, V. and A. Hanzlikova. 1972. Root exudates of plants. IV. Differences in chemical composition of seed and seedlings exudates. Plant and Soil 36:271-282.

Vaughan, D. and B.G. Ord. 1991. Extraction of potential allelochemicals and their effects on root morphology and nutrient contents. In: Plant root growth: an ecological perspective. Atkinson, D. Ed. Blackwell Scientific Publications. London. 399-421.

Veeken, A.H.M., and H.V.M. Hamelers. 1999. Removal of heavy metals from sewage sludge by extraction with organic acids. Wat. Sci. Technol. 40:129-136.

Vempati, R.K., K.P. Kollipara, J.W. Stucki, and H. Wilkinson. 1995. Reduction of structural iron in selected iron-bearing minerals by soybean root exudates grown in an in vitro geoponic system. J. Plant Nutr. 18:343-353.

Vempati, R.K. and R.H. Loeppert. 1988. Chemistry and mineralogy of Fe-containing oxides and layer silicates in relation to plant available iron. Journal of Plant Nutrition 11:1557-1574.

Visser, J.H., R. Herb, and H. Schildknecht. 1987. Recovery and preliminary chromatographic investigation of germination stimulants produced by *Vigna unguiculata* Walp. cv Saunders Upright. J. Plant Physiol. 129:375-381.

Wasay, S.A., S.F. Barrington, and S. Tokunaga. 1998. Remediation of soils polluted by heavy metals using salts of organic acids and chelating agents. Environ. Technol. 19:369-380.

Waschutza, S., N. Hofmann, E.G. Niemann, and I. Fendrik. 1992. Investigations on root exudates of Korean rice. Symbiosis 13:181-189.

Welch, R.M. 1995. Micronutrient nutrition of plants. Critical Reviews in Plant Sciences 14:49-82.

Wiedenroth, E. and J. Poskuta. 1981. The influence of oxygen deficiency in roots on CO_2 carbon dioxide exchange rates of shoots and distribution of ^{14}C carbon isotope-photoassimilates of wheat seedlings. Z. Pflanzenphysiol. 103:459-467.

Wittenmayer, L., and K. Szabo. 2000. The role of root exudates in specific apple (*Malus domestica* Borkh.) replant disease (SARD). J. Plant Nutr. Soil Sci. 163:399-404.

Yang, C-H., and D.E. Crowley. 2000. Rhizosphere microbial community structure in relation to root location and plant iron nutritional status. Appl. Environ. Microbiol. 66:345-351.

Young, C.C., C.H. Chang, L.F. Chen, and C.C. Chao. 1998. Characterization of the nitrogen fixation and ferric phosphate solubilizing bacteria isolated from a Taiwan soil. J. of the Chinese Agricultural Chemical Society 36:201-210.

Youssef, R.A., and M. Chino. 1989. Root-induced changes in the rhizosphere of plants. Soil Sci. Plant Nutrition. 35:461-468.

Yu, J.Q. and Y. Matsui. 1994. Phytotoxic substances in root exudates of cucumber (*Cucumis sativus* L.). J. Chem. Ecol. 20:21-31.

Yu, J.Q. and Y. Matsui. 1997. Effects of root exudates of cucumber (*Cucumis sativus*) and allelochemicals on ion uptake by cucumber seedlings. J. Chem. Ecol. 23:817-827.

Zhang, F., V. Romheld, and H. Marschner. 1989. Effect of zinc deficiency in wheat on the release of zinc and iron mobilizing root exudates. Zeitschrift fur Pflanzenernahrung und Bodenkunde. 152:205-210.

Zhang, F., V. Romheld, and H. Marschner. 1991. Release of zinc mobilizing root exudates in different plant species as affected by zinc nutritional status. Journal of Plant Nutrition 14:675-686.

Zhang, T.C., and H. Pang. 1999. Applications of microelectrode techniques to measure pH and oxidation-reduction potential in rhizosphere soil. Environ. Sci. Technol. 33:1293-1299.

WASTEWATER UTILITY

Alabama
Montgomery Water Works & Sanitary Sewer Board

Alaska
Anchorage Water & Wastewater Utility

Arizona
Gila Resources
Glendale, City of, Utilities Department
Mesa, City of
Peoria, City of
Phoenix Water Services Department
Pima County Wastewater Management

Arkansas
Little Rock Wastewater Utility

California
Calaveras County Water District
Central Contra Costa Sanitary District
Contra Costa Water District
Crestline Sanitation District
Delta Diablo Sanitation District
Dublin San Ramon Services District
East Bay Dischargers Authority
East Bay Municipal Utility District
El Dorado Irrigation District
Fairfield-Suisun Sewer District
Fresno Department of Public Utilities
Irvine Ranch Water District
Las Virgenes Municipal Water District
Livermore, City of
Lodi, City of
Los Angeles, City of
Los Angeles County, Sanitation Districts of
Napa Sanitation District
Orange County Sanitation District
Palo Alto, City of
Riverside, City of
Sacramento Regional County Sanitation District
San Diego Metropolitan Wastewater Department, City of
San Francisco, City & County of
San Jose, City of
Santa Barbara, City of
Santa Rosa, City of
South Bayside System Authority
South Coast Water District
South Orange County Wastewater Authority
Union Sanitary District
West Valley Sanitation District

Colorado
Boulder, City of
Colorado Springs, City of
Littleton/Englewood Water Pollution Control Plant
Metro Wastewater Reclamation District, Denver

Connecticut
The Mattabasset District
New Haven, City of, WPCA

District of Columbia
District of Columbia Water & Sewer Authority

Florida
Broward, County of
Fort Lauderdale, City of
Gainesville Regional Utilities
JEA
Miami-Dade Water & Sewer Authority
Orange County Utilities Department
Orlando, City of
Reedy Creek Improvement District
Seminole County Environmental Services
St. Petersburg, City of
Stuart Public Utilities
Tallahassee, City of
Tampa, City of
Toho Water Authority
West Palm Beach, City of

Georgia
Atlanta Department of Watershed Management
Augusta, City of
Clayton County Water Authority
Cobb County Water System
Columbus Water Works
Fulton County
Gwinnett County Department of Public Utilities
Macon Water Authority
Savannah, City of

Hawaii
Honolulu, City and County of

Illinois
American Bottoms Wastewater Treatment Plant
Greater Peoria Sanitary District
Kankakee River Metropolitan Agency
Metropolitan Water Reclamation District of Greater Chicago
North Shore Sanitary District
Wheaton Sanitary District

Indiana
Fort Wayne, City of

Iowa
Ames, City of
Cedar Rapids Wastewater Facility
Des Moines Metro Wastewater Reclamation Authority
Iowa City, City of

Kansas
Johnson County Unified Wastewater Districts
Unified Government of Wyandotte County/Kansas City, City of

Kentucky
Louisville & Jefferson County Metropolitan Sewer District

Louisiana
Sewerage & Water Board of New Orleans

Maine
Bangor, City of
Portland Water District

Maryland
Anne Arundel County Bureau of Utility Operations
Howard County Department of Public Works
Washington Suburban Sanitary Commission

Massachusetts
Boston Water & Sewer Commission
Upper Blackstone Water Pollution Abatement District

Michigan
Ann Arbor, City of
Detroit, City of
Holland Board of Public Works
Lansing, City of
Owosso Mid-Shiawassee County WWTP
Saginaw, City of
Wayne County Department of Environment
Wyoming, City of

Minnesota
Rochester, City of
Western Lake Superior Sanitary District

Missouri
Independence, City of
Kansas City Missouri Water Services Department
Little Blue Valley Sewer District

Nebraska
Lincoln Wastewater System

Nevada
Henderson, City of

New Jersey
Passaic Valley Sewerage Commissioners

New York
New York City Department of Environmental Protection
Rockland County Solid Waste Management Authority/ Sewer District

North Carolina
Cary, Town of
Charlotte/Mecklenburg Utilities
Durham, City of
Metropolitan Sewerage District of Buncombe County
Orange Water & Sewer Authority

Ohio
Akron, City of
Butler County Department of Environmental Services
Columbus, City of
Metropolitan Sewer District of Greater Cincinnati
Northeast Ohio Regional Sewer District
Summit, County of

Oklahoma
Tulsa, City of

Oregon
Clean Water Services
Eugene/Springfield Water Pollution Control
Water Environment Services

Pennsylvania
Philadelphia, City of
University Area Joint Authority, State College

South Carolina
Charleston Commissioners of Public Works
Mount Pleasant Waterworks & Sewer Commission
Spartanburg Sanitary Sewer District

Tennessee
Cleveland, City of
Knoxville Utilities Board
Murfreesboro Water & Sewer Department
Nashville Metro Water Services

Texas
Austin, City of
Dallas Water Utilities
Denton, City of
El Paso Water Utilities
Fort Worth, City of
Gulf Coast Waste Disposal Authority

Houston, City of

San Antonio Water System

Trinity River Authority

Utah
Salt Lake City Corporation

Virginia
Alexandria Sanitation Authority

Arlington, County of

Fairfax County Virginia

Hampton Roads Sanitation District

Henrico, County of

Hopewell Regional Wastewater Treatment Facility

Loudoun County Sanitation Authority

Lynchburg Regional WWTP

Prince William County Service Authority

Richmond, City of

Rivanna Water & Sewer Authority

Washington
Edmonds, City of

Everett, City of

King County Department of Natural Resources

Seattle Public Utilities

Sunnyside, Port of

Yakima, City of

Wisconsin
Green Bay Metro Sewerage District

Madison Metropolitan Sewerage District

Milwaukee Metropolitan Sewerage District

Racine, City of

Sheboygan Regional Wastewater Treatment

Wausau Water Works

Australia
South Australian Water Corp.

Sydney Water Corp.

Water Corp. of Western Australia

Canada
Greater Vancouver Regional District

Toronto, City of, Ontario

Winnipeg, City of, Manitoba

Mexico
Servicios de Agua y Drenaje de Monterrey, I.P.D.

New Zealand
Watercare Services Limited

Singapore
Singapore Public Utilities Board

United Kingdom
Yorkshire Water Services Limited

STORMWATER UTILITY

California
Los Angeles, City of, Department of Public Works

Monterey, City of

Santa Rosa, City of

Colorado
Boulder, City of

Georgia
Griffin, City of

Iowa
Cedar Rapids Wastewater Facility

Des Moines Metro Wastewater Reclamation Authority

Kansas
Overland Park, City of

Kentucky
Louisville & Jefferson County Metropolitan Sewer District

Maine
Portland Water District

Minnesota
Western Lake Superior Sanitary District

North Carolina
Cary, Town of

Charlotte, City of, Stormwater Services

Pennsylvania
Philadelphia, City of

Tennessee
Chattanooga Storm Water Management

Washington
Seattle Public Utilities

STATE

Arkansas Department of Environmental Quality

Fresno Metropolitan Flood Control District, Calif.

Urban Drainage & Flood Control District, Colo.

CORPORATE

ADS Environmental Services

The ADVENT Group Inc.

Alan Plummer & Associates

Alden Research Laboratory

Alpine Technology Inc.

Aqua-Aerobic Systems, Inc.

Aquateam–Norwegian Water Technology Centre A/S

BaySaver Inc.

BioVir Laboratories, Inc.

Black & Veatch

Boyle Engineering Corporation

BPR CSO

Brown & Caldwell

Burns & McDonnell

CABE Associates Inc.

The Cadmus Group

Camp Dresser & McKee Inc.

Carollo Engineers Inc.

Carpenter Environmental Associates Inc.

CDS Technologies Inc.

Chemtrac Systems Inc.

CH2M HILL

Damon S. Williams Associates, LLC

David L. Sheridan, P.C.

Earth Tech Inc.

Eco-Matrix

Environmental Engineers International

EMA Inc.

The Eshelman Company Inc.

Finkbeiner, Pettis, & Strout Inc.

Freese & Nichols, Inc.

Frontier Geosciences Inc.

ftn Associates Inc.

Fuss & O'Neill, Inc.

Gannett Fleming Inc.

Golder Associates Inc.

Greeley and Hansen LLC

Hazen & Sawyer, P.C.

HDR Engineering Inc.

HNTB Corporation

HydroQual Inc.

Ingersoll-Rand Energy Systems

Insituform Technologies Inc.

Institute for Environmental Technology & Industry, Korea

Jacobson Helgoth Consultants Inc.

Jason Consultants Inc.

Jordan, Jones, & Goulding Inc.

KCI Technologies Inc.

Kelly & Weaver, P.C.

Kennedy/Jenks Consultants

Komline Sanderson Engineering Corporation

Lawler, Matusky & Skelly Engineers, LLP

Limno-Tech Inc.

Lombardo Associates Inc.

Malcolm Pirnie Inc.

McKim & Creed

MEC Analytical Systems Inc.

Metcalf & Eddy Inc.

MWH

New England Organics

Odor & Corrosion Technology Consultants Inc. (OCTC)

ONDEO Degremont Inc.

Oswald Green, LLC

PA Government Services Inc.

Parametrix Inc.

Parsons

Post, Buckley, Schuh & Jernigan

R&D Engineering/Conestoga Rover & Associates

The RETEC Group

RMC, Inc. (Raines, Melton & Carella)

R.M. Towill Corporation

Ross & Associates Ltd.

Royce Technologies

SAIC Maritime Technical Group

Sear-Brown

Short Elliott Hendrickson, Inc.

Stormwater Management, Inc.

Synagro Technologies, Inc.

Tetra Tech Inc.

Trojan Technologies Inc.

URS Corporation

Veolia Water NATC

Wade-Trim Inc.

Weston Solutions Inc.

Woodard & Curran

WRc/D&B, LLC

WWETCO, LLC

Zoeller Pump Company

INDUSTRY

American Electric Power

ChevronTexaco Energy Research & Technology Company

The Coca-Cola Company

Dow Chemical Company

DuPont Company

Eastman Kodak Company

Merck & Company Inc.

ONDEO Services

Procter & Gamble Company

PSEG Services Corp.

Severn Trent Services Inc.

Shell Global Solutions

Thames Water Plc

United Water Services LLC

Note: List as of 1/31/04

WERF Product Order Form

As a benefit of joining the Water Environment Research Foundation, subscribers are entitled to receive one complimentary copy of all final reports and other products.Additional copies are available at cost (usually $10).To order your complimentary copy of a report, please write "free" in the unit price column.WERF keeps track of all orders.If the charge differs from what is shown here, we will call to confirm the total before processing.

Name ____________________ Title ____________________

Organization ____________________

Address ____________________

City ____________ State ______ Zip Code ______ Country ____________

Phone ____________ Fax ____________ Email ____________

Stock #	Product	Quantity	Unit Price	Total

Postage & Handling	
VA Residents Add 4.5% Sales Tax	
Canadian Residents Add 7% GST	
TOTAL	

Method of Payment:

(All orders must be prepaid.)

- ❑ Check or Money Order Enclosed
- ❑ Visa ❑ Mastercard ❑ American Express

Account No. ____________________ Exp. Date ________

Signature ____________________

Shipping & Handling:

Amount of Order	United States	Canada & Mexico	All Others
Up to but not more than:	Add:	Add:	Add:
$20.00	$5.00*	$8.00	50% of amount
30.00	5.50	8.00	40% of amount
40.00	6.00	8.00	
50.00	6.50	14.00	
60.00	7.00	14.00	
80.00	8.00	14.00	
100.00	10.00	21.00	
150.00	12.50	28.00	
200.00	15.00	35.00	
More than $200.00	Add 20% of order	Add 20% of order	

*minimum amount for all orders

Note: Please make checks payable to the Water Environment Research Foundation.

To Order (Subscribers Only):

Log on to www.werf.org and click on the "Product Catalog."

Phone: (703) 684-2470
Fax: (703) 299-0742.

WERF
Attn: Subscriber Services
635 Slaters Lane
Alexandria, VA 22314-1177

To Order (Non-Subscribers):

Non-subscribers may be able to order WERF publications either through WEF (www.wef.org) or IWAP (www.iwapublishing.com).Visit WERF's website at www.werf.org for details.

IWA Publishing's authorised EU representative for General Product Safety Regulations is Diane D'Arras, 15 rue Duret, 75116 Paris, France, e-mail: safety@iwap.co.uk.

Printed and bound by CPI Group (UK) Ltd, Croydon, CR0 4YY
12/05/2026
02108886-0006